The Gravitational Wave

" Ripples In Space-Time "

Edited by Paul F. Kisak

Contents

1 Gravitational wave   **1**

  1.1 Introduction 1

  1.2 Effects of a passing gravitational wave 3

  1.3 Sources of gravitational waves 6

    1.3.1 Power radiated by orbiting bodies 6

    1.3.2 Orbital decay from gravitational radiation 7

    1.3.3 Orbital lifetime limits from gravitational radiation 9

    1.3.4 Wave amplitudes from the Earth–Sun system 10

    1.3.5 Radiation from other sources 11

  1.4 Astrophysics and gravitational waves 11

    1.4.1 Energy, momentum, and angular momentum carried by gravitational waves 12

  1.5 Detecting gravitational waves 13

    1.5.1 Difficulties in detection 13

    1.5.2 Ground-based interferometers 14

    1.5.3 Using pulsar timing arrays 15

    1.5.4 Einstein@Home 16

    1.5.5 Primordial gravitational waves 16

  1.6 Mathematics 16

    1.6.1 Linear approximation 17

    1.6.2 Relation to the source 18

  1.7 See also 19

  1.8 References 20

  1.9 Further reading 22

  1.10 Bibliography 23

  1.11 External links 23

2 Gravitational field   **24**

  2.1 Classical mechanics 24

  2.2 General relativity 25

  2.3 See also 25

2.4   Notes   26

3   Gravitational wave background   **27**

3.1   See also   27

3.2   External links   27

3.3   References   27

4   Cosmic microwave background   **28**

4.1   Features   29

4.2   History   30

4.2.1   Timeline   32

4.3   Relationship to the Big Bang   34

4.3.1   Primary anisotropy   35

4.3.2   Late time anisotropy   36

4.4   Polarization   37

4.4.1   E-modes   37

4.4.2   B-modes   37

4.5   Microwave background observations   39

4.6   Data reduction and analysis   40

4.6.1   CMBR dipole anisotropy   41

4.6.2   Low multipoles and other anomalies   41

4.7   Future evolution   41

4.8   In popular culture   42

4.9   See also   42

4.10 References   42

4.11 External links   48

5   Graviton   **49**

5.1   Theory   49

5.1.1   Gravitons and renormalization   49

5.1.2   Comparison with other forces   49

5.1.3   Gravitons in speculative theories   50

5.2   Experimental observation   50

5.3   Difficulties and outstanding issues   50

5.4   See also   51

5.5   References   51

5.6   External links   52

6   Gravitoelectromagnetism   **53**

6.1   Background   54

6.2 Equations 54

    6.2.1 Lorentz force 55

    6.2.2 Poynting vector 55

    6.2.3 Scaling of fields 55

    6.2.4 In Planck units 55

6.3 Higher-order effects 55

6.4 Gravitomagnetic fields of astronomical objects 56

    6.4.1 Earth 56

    6.4.2 Pulsar 57

6.5 Lack of invariance 57

6.6 See also 57

6.7 References 57

6.8 Further reading 58

    6.8.1 Books 58

    6.8.2 Papers 59

6.9 External links 59

7 Gravitational-wave observatory   **60**

7.1 Complications 61

7.2 Weber bars 61

7.3 Interferometers 61

7.4 High frequency detectors 62

7.5 Pulsar timing arrays 63

7.6 Einstein@Home 63

7.7 Specific operational and planned gravitational-wave detectors 63

7.8 References 64

8 LIGO   **66**

8.1 Mission 66

8.2 Observatories 67

8.3 Operation 68

8.4 Observations 69

    8.4.1 Enhanced LIGO 70

8.5 Future 70

    8.5.1 Advanced LIGO 70

    8.5.2 LIGO-India 70

8.6 See also 71

8.7 Notes 71

8.8 References 72

        8.8.1   Further reading   72

    8.9   External links   73

9   Virgo interferometer   **74**

    9.1   Description   74

    9.2   History   75

    9.3   References   75

    9.4   External links   76

10 Evolved Laser Interferometer Space Antenna   **77**

    10.1 Mission description   77

    10.2 LISA Pathfinder   78

    10.3 Science   78

    10.4 Other gravitational-wave experiments   79

    10.5 History   80

    10.6 See also   80

    10.7 External links   80

    10.8 References   80

11 Linearized gravity   **83**

    11.1 The method   83

    11.2 Applications   83

        11.2.1 Weak-field approximation   83

    11.3 Linearised Einstein field equations   84

        11.3.1 Derivation for the Minkowski metric   84

    11.4 With a coordinate condition   85

    11.5 Applications   86

    11.6 See also   86

    11.7 References   86

12 Quadrupole formula   **87**

    12.1 References   87

13 SQUID   **88**

    13.1 History and design   89

        13.1.1 DC SQUID   89

        13.1.2 RF SQUID   91

        13.1.3 Materials used   91

    13.2 Uses   91

        13.2.1 Proposed uses   93

13.3 See also 93

13.4 Notes 94

13.5 References 94

14 MiniGrail   **95**

14.1 References 95

14.2 External links 96

15 GEO600   **97**

15.1 Hardware 97

    15.1.1 Suspensions and seismic isolation 97

    15.1.2 Optics 97

    15.1.3 Advanced features 97

15.2 Sensitivity and measurements 98

15.3 Data/ Einstein@home 98

15.4 Joint science run with LIGO 98

15.5 Claimed link between GEO600 detector noise and holographic properties of spacetime 99

15.6 See also 99

15.7 References 99

15.8 External links 100

16 TAMA 300   **101**

16.1 External links 102

17 KAGRA   **103**

17.1 References 103

17.2 External 103

18 Deci-hertz Interferometer Gravitational wave Observatory   **104**

18.1 See also 104

18.2 References 104

19 pp-wave spacetime   **105**

19.1 Mathematical definition 105

19.2 Physical interpretation 106

19.3 Relation to other classes of exact solutions 107

19.4 Relation to other theories 108

19.5 Geometric and physical properties 108

19.6 Examples 109

19.7 See also 109

19.8 References 110

19.9 External links 110

20 Spin-flip   **111**

20.1 Physics of Spin-Flips 111

20.2 Connection with radio galaxies 112

20.3 See also 112

20.4 References 112

20.5 External links 113

21 Sticky bead argument   **114**

21.1 Description of the thought experiment 114

21.2 History of arguments on the properties of gravitational waves 114

    21.2.1 Einstein's double reversal 114

    21.2.2 The Bern and Chapel Hill conferences 115

    21.2.3 Feynman's argument 116

    21.2.4 Rosen's final views 116

21.3 See also 116

21.4 Notes 116

21.5 References 117

21.6 Text and image sources, contributors, and licenses 118

    21.6.1 Text 118

    21.6.2 Images 121

    21.6.3 Content license 123

Chapter 1

Gravitational wave

Not to be confused with gravity wave.

In physics, **gravitational waves** are ripples in the curvature of spacetime which propagate as waves, travelling outward from the source. Predicted in 1916[1][2] by Albert Einstein to exist on the basis of his theory of general relativity,[3][4] gravitational waves theoretically transport energy as **gravitational radiation**. Sources of detectable gravitational waves could possibly include binary star systems composed of white dwarfs, neutron stars, or black holes. The existence of gravitational waves is a possible consequence of the Lorentz invariance of general relativity since it brings the concept of a limiting speed of propagation of the physical interactions with it. Gravitational waves cannot exist in the Newtonian theory of gravitation, in which physical interactions propagate at infinite speed.

Although gravitational radiation has not been *directly* detected, there is *indirect* evidence for its existence.[5] For example, the 1993 Nobel Prize in Physics was awarded for measurements of the Hulse–Taylor binary system which suggest that gravitational waves are more than mathematical anomalies. Various gravitational wave detectors exist and on 17 March 2014, astronomers at the Harvard–Smithsonian Center for Astrophysics erroneously claimed that they had detected and produced "the first direct image of gravitational waves across the primordial sky" within the cosmic microwave background, providing flawed evidence for inflation and the Big Bang.[5][6][7][8][9][10] On 19 June 2014, lowered confidence in confirming the cosmic inflation findings was reported[11][12][13] and on 19 September 2014, a further reduction in confidence was reported.[14][15] On 30 January 2015, even less confidence yet was reported;[16][17] Nature went as far as publishing a news article entitled "Gravitational waves discovery now officially dead".[18]

1.1 Introduction

In Einstein's theory of general relativity, gravity is treated as a phenomenon resulting from the curvature of spacetime. This curvature is caused by the presence of mass. Generally, the more mass that is contained within a given volume of space, the greater the curvature of spacetime will be at the boundary of this volume.[5] As objects with mass move around in spacetime, the curvature changes to reflect the changed locations of those objects. In certain circumstances, accelerating objects generate changes in this curvature, which propagate outwards at the speed of light in a wave-like manner. These propagating phenomena are known as gravitational waves.

As a gravitational wave passes a distant observer, that observer will find spacetime distorted by the effects of strain. Distances between free objects increase and decrease rhythmically as the wave passes, at a frequency corresponding to that of the wave. This occurs despite such free objects never being subjected to an unbalanced force. The magnitude of this effect decreases inversely with distance from the source. Inspiralling binary neutron stars are predicted to be a powerful source of gravitational waves as they coalesce, due to the very large acceleration of their masses as they orbit close to one another. However, due to the astronomical distances to these sources the effects when measured on Earth are predicted to be very small, having strains of less than 1 part in 10^{20}. Scientists are attempting to demonstrate the existence of these waves with ever more sensitive detectors. The current most sensitive measurement is about one part in $5{\times}10^{22}$

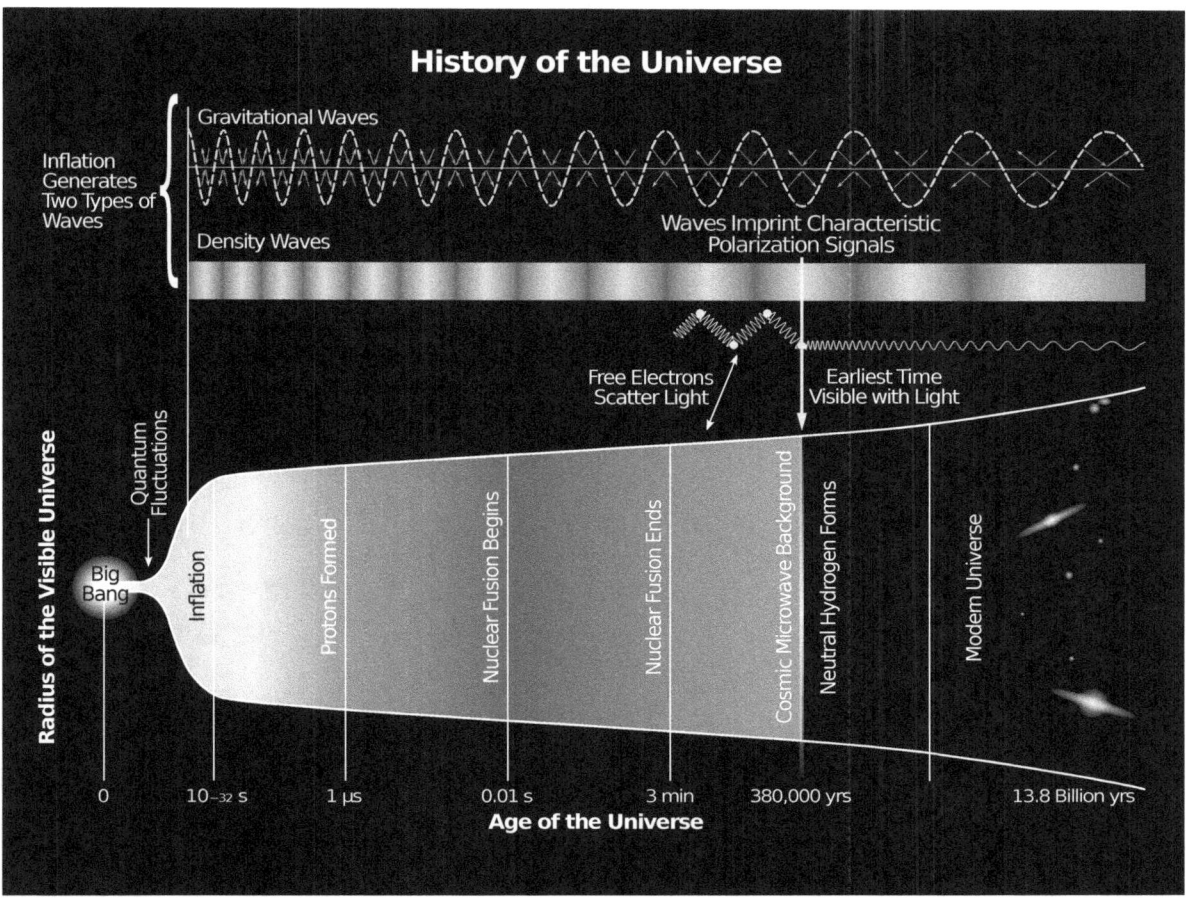

History of the Universe - gravitational waves are hypothesized to arise from cosmic inflation, a faster-than-light expansion just after the Big Bang (17 March 2014).[7][9][10]

(as of 2012) provided by the LIGO and VIRGO observatories.[19] The lack of detection in these observatories provides an upper limit on the frequency of such powerful sources.[20][21] A space based observatory, the Laser Interferometer Space Antenna, is currently under development by ESA.

Gravitational waves should penetrate regions of space that electromagnetic waves cannot. It is hypothesized that they will be able to provide observers on Earth with information about black holes and other exotic objects in the distant Universe. Such systems cannot be observed with more traditional means such as optical telescopes and radio telescopes. In particular, gravitational waves could be of interest to cosmologists as they offer a possible way of observing the very early universe. This is not possible with conventional astronomy, since before recombination the universe was opaque to electromagnetic radiation.[22] Precise measurements of gravitational waves will also allow scientists to test the general theory of relativity more thoroughly.

In principle, gravitational waves could exist at any frequency. However, very low frequency waves would be impossible to detect and there is no credible source for detectable waves of very high frequency. Stephen W. Hawking and Werner Israel list different frequency bands for gravitational waves that could be plausibly detected, ranging from 10^{-7} Hz up to 10^{11} Hz.[23]

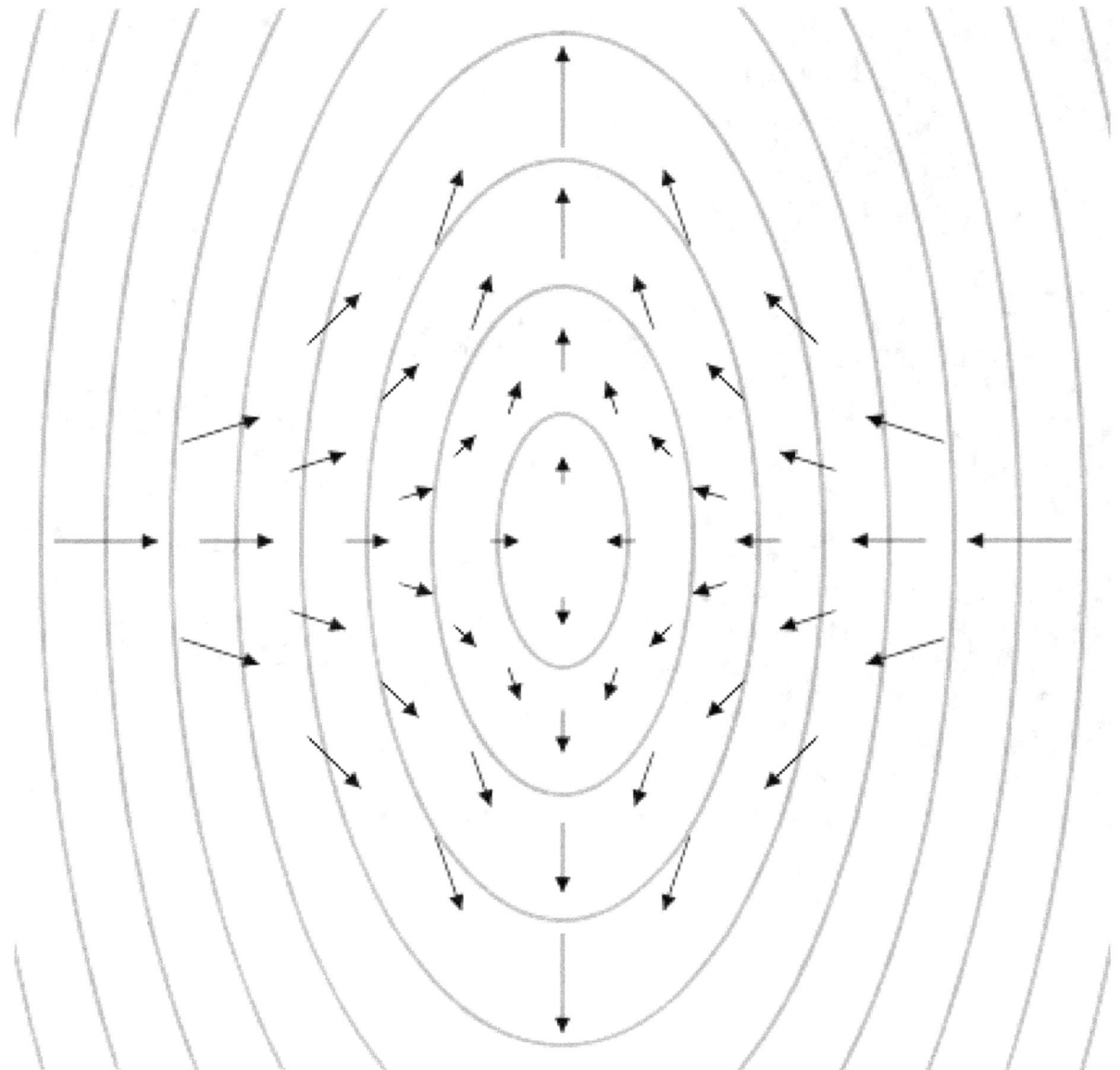

Linearly polarised gravitational wave

1.2 Effects of a passing gravitational wave

The effects of a passing gravitational wave can be visualized by imagining a perfectly flat region of spacetime with a group of motionless test particles lying in a plane (the surface of your screen). As a gravitational wave passes through the particles along a line perpendicular to the plane of the particles (i.e. following your line of vision into the screen), the particles will follow the distortion in spacetime, oscillating in a "cruciform" manner, as shown in the animations. The area enclosed by the test particles does not change and there is no motion along the direction of propagation.

The oscillations depicted here in the animation are exaggerated for the purpose of discussion—in reality a gravitational wave has a very small amplitude (as formulated in linearized gravity). However, they enable us to visualize the kind of oscillations associated with gravitational waves as produced for example by a pair of masses in a circular orbit. In this case the amplitude of the gravitational wave is constant, but its plane of polarization changes or rotates at twice the orbital rate and so the time-varying gravitational wave size (or 'periodic spacetime strain') exhibits a variation as shown in the

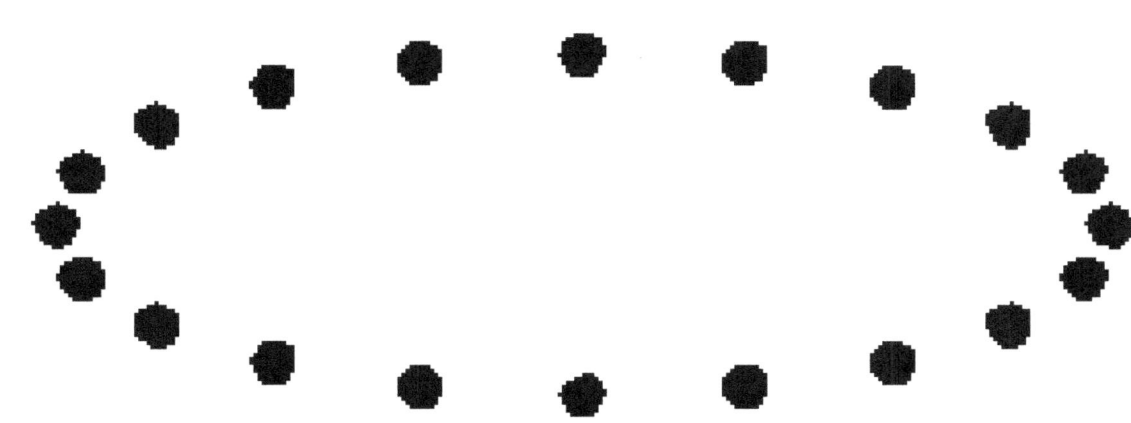

The effect of a plus-polarized gravitational wave on a ring of particles.

animation.[24] If the orbit is elliptical then the gravitational wave's amplitude also varies with time according to Einstein's quadrupole formula.[25]

Like other waves, there are a few useful characteristics describing a gravitational wave:

- **Amplitude**: Usually denoted h , this is the size of the wave — the fraction of stretching or squeezing in the animation. The amplitude shown here is roughly $h = 0.5$ (or 50%). Gravitational waves passing through the Earth are many billions times weaker than this — $h \approx 10^{-20}$. Note that this is not the quantity that would be analogous to what is usually called the amplitude of an electromagnetic wave, which would be $\frac{dh}{dt}$.

- **Frequency**: Usually denoted f, this is the frequency with which the wave oscillates (1 divided by the amount of time between two successive maximum stretches or squeezes)

- **Wavelength**: Usually denoted λ , this is the distance along the wave between points of maximum stretch or squeeze.

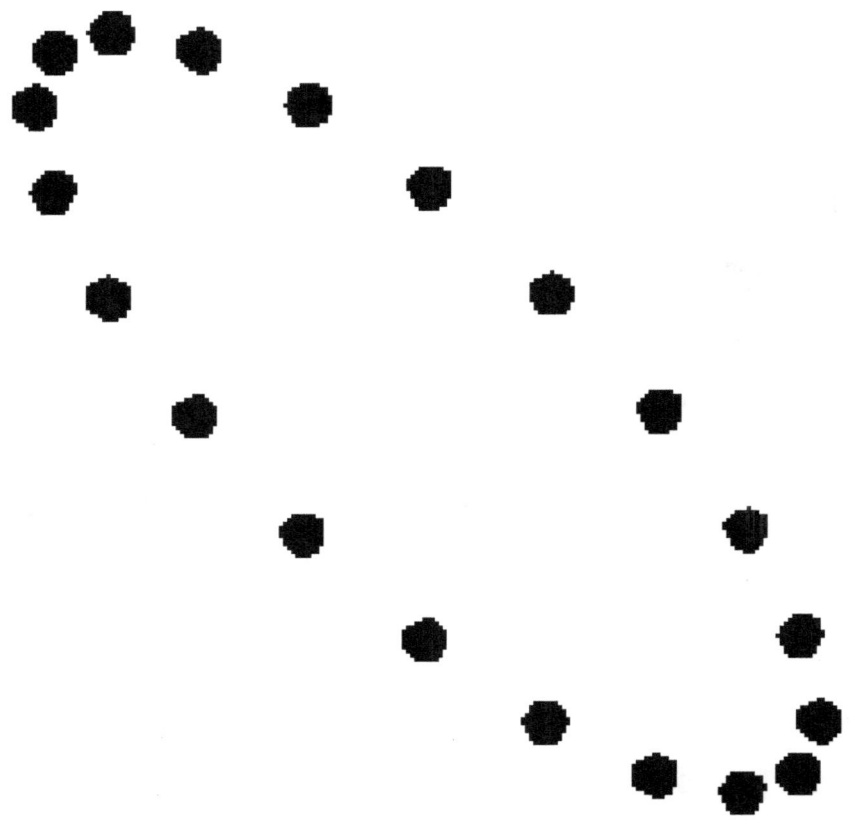

The effect of a cross-polarized gravitational wave on a ring of particles.

- **Speed**: This is the speed at which a point on the wave (for example, a point of maximum stretch or squeeze) travels. For gravitational waves with small amplitudes, this is equal to the speed of light, c .

The speed, wavelength, and frequency of a gravitational wave are related by the equation $c = \lambda f$, just like the equation for a light wave. For example, the animations shown here oscillate roughly once every two seconds. This would correspond to a frequency of 0.5 Hz, and a wavelength of about 600,000 km, or 47 times the diameter of the Earth.

In the example just discussed, we actually assume something special about the wave. We have assumed that the wave is linearly polarized, with a "plus" polarization, written h_+ . Polarization of a gravitational wave is just like polarization of a light wave except that the polarizations of a gravitational wave are at 45 degrees, as opposed to 90 degrees. In particular, if we had a "cross"-polarized gravitational wave, $h_\times$, the effect on the test particles would be basically the same, but rotated by 45 degrees, as shown in the second animation. Just as with light polarization, the polarizations of gravitational waves may also be expressed in terms of circularly polarized waves. Gravitational waves are polarized because of the nature of their sources. The polarization of a wave depends on the angle from the source, as we will see in the next section.

1.3   Sources of gravitational waves

In general terms, gravitational waves are radiated by objects whose motion involves acceleration, provided that the motion is not perfectly spherically symmetric (like an expanding or contracting sphere) or cylindrically symmetric (like a spinning disk or sphere). A simple example of this principle is provided by the spinning dumbbell. If the dumbbell spins like wheels on an axle, it will not radiate gravitational waves; if it tumbles end over end like two planets orbiting each other, it will radiate gravitational waves. The heavier the dumbbell, and the faster it tumbles, the greater is the gravitational radiation it will give off. If we imagine an extreme case in which the two weights of the dumbbell are massive stars like neutron stars or black holes, orbiting each other quickly, then significant amounts of gravitational radiation would be given off.

Some more detailed examples:

- Two objects orbiting each other in a quasi-Keplerian planar orbit (basically, as a planet would orbit the Sun) *will* radiate.

- A spinning non-axisymmetric planetoid — say with a large bump or dimple on the equator — *will* radiate.

- A supernova *will* radiate except in the unlikely event that the explosion is perfectly symmetric.

- An isolated non-spinning solid object moving at a constant velocity *will not* radiate. This can be regarded as a consequence of the principle of conservation of linear momentum.

- A spinning disk *will not* radiate. This can be regarded as a consequence of the principle of conservation of angular momentum. However, it *will* show gravitomagnetic effects.

- A spherically pulsating spherical star (non-zero monopole moment or mass, but zero quadrupole moment) *will not* radiate, in agreement with Birkhoff's theorem.

More technically, the third time derivative of the quadrupole moment (or the *l*-th time derivative of the *l*-th multipole moment) of an isolated system's stress–energy tensor must be nonzero in order for it to emit gravitational radiation. This is analogous to the changing dipole moment of charge or current necessary for electromagnetic radiation.

1.3.1   Power radiated by orbiting bodies

Gravitational waves carry energy away from their sources and, in the case of orbiting bodies, this is associated with an inspiral or decrease in orbit. Imagine for example a simple system of two masses — such as the Earth-Sun system — moving slowly compared to the speed of light in circular orbits. Assume that these two masses orbit each other in a circular orbit in the $x - y$ plane. To a good approximation, the masses follow simple Keplerian orbits. However, such an orbit represents a changing quadrupole moment. That is, the system will give off gravitational waves.

Suppose that the two masses are m_1 and m_2, and they are separated by a distance r. The power given off (radiated) by this system is:

$$P = \frac{dE}{dt} = -\frac{32}{5} \frac{G^4}{c^5} \frac{(m_1 m_2)^2 (m_1 + m_2)}{r^5} \text{, [26]}$$

where G is the gravitational constant, c is the speed of light in vacuum and where the negative sign means that power is being given off by the system, rather than received. For a system like the Sun and Earth, r is about 1.5×10^{11} m and m_1 and m_2 are about 2×10^{30} and 6×10^{24} kg respectively. In this case, the power is about 200 watts. This is truly tiny compared to the total electromagnetic radiation given off by the Sun (roughly 3.86×10^{26} watts).

In theory, the loss of energy through gravitational radiation could eventually drop the Earth into the Sun. However, the total energy of the Earth orbiting the Sun (kinetic energy + gravitational potential energy) is about 1.14×10^{36} joules of which only 200 joules per second is lost through gravitational radiation, leading to a decay in the orbit by about 1×10^{-15} meters per day or roughly the diameter of a proton. At this rate, it would take the Earth approximately 1×10^{13} times more than the current age of the Universe to spiral onto the Sun. This estimate overlooks the decrease in r over time, but the majority of the time the bodies are far apart and only radiating slowly, so the difference is unimportant in this example.

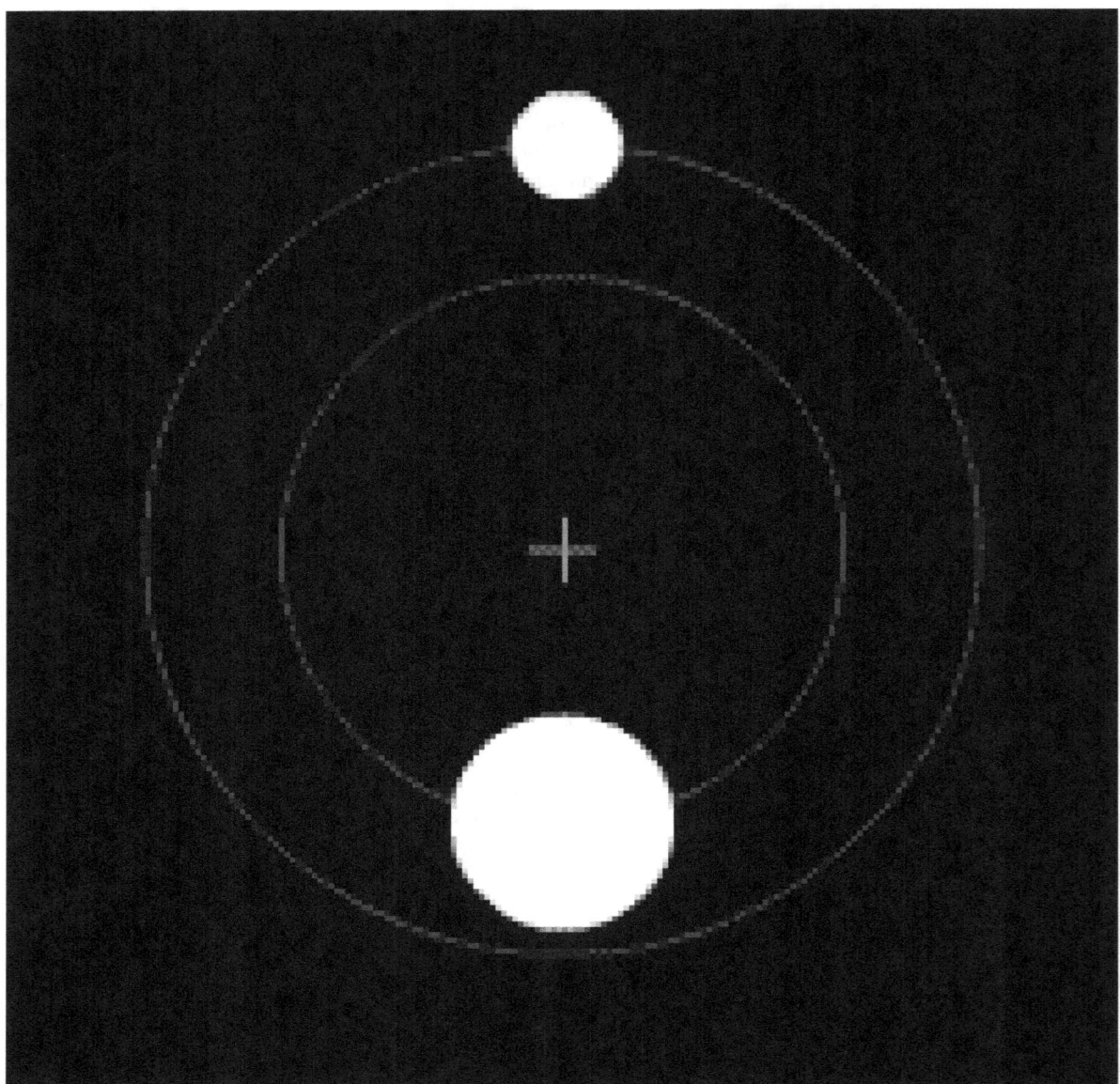

Two stars of dissimilar mass are in circular orbits. Each revolves about their common center of mass (denoted by the small red cross) in a circle with the larger mass having the smaller orbit.

A more dramatic example of radiated gravitational energy is represented by two solar mass ($M\odot$) neutron stars orbiting at a distance from each other of 1.89×10^8 m (only 0.63 light-seconds apart). [The Sun is 8 light minutes from the Earth.] Plugging their masses into the above equation shows that the gravitational radiation from them would be 1.38×10^{28} watts, which is about 100 times more than the Sun's electromagnetic radiation.

1.3.2 Orbital decay from gravitational radiation

See also: Two-body problem in general relativity

Gravitational radiation robs the orbiting bodies of energy. It first circularizes their orbits and then gradually shrinks their radius. As the energy of the orbit is reduced, the distance between the bodies decreases, and they rotate more rapidly.

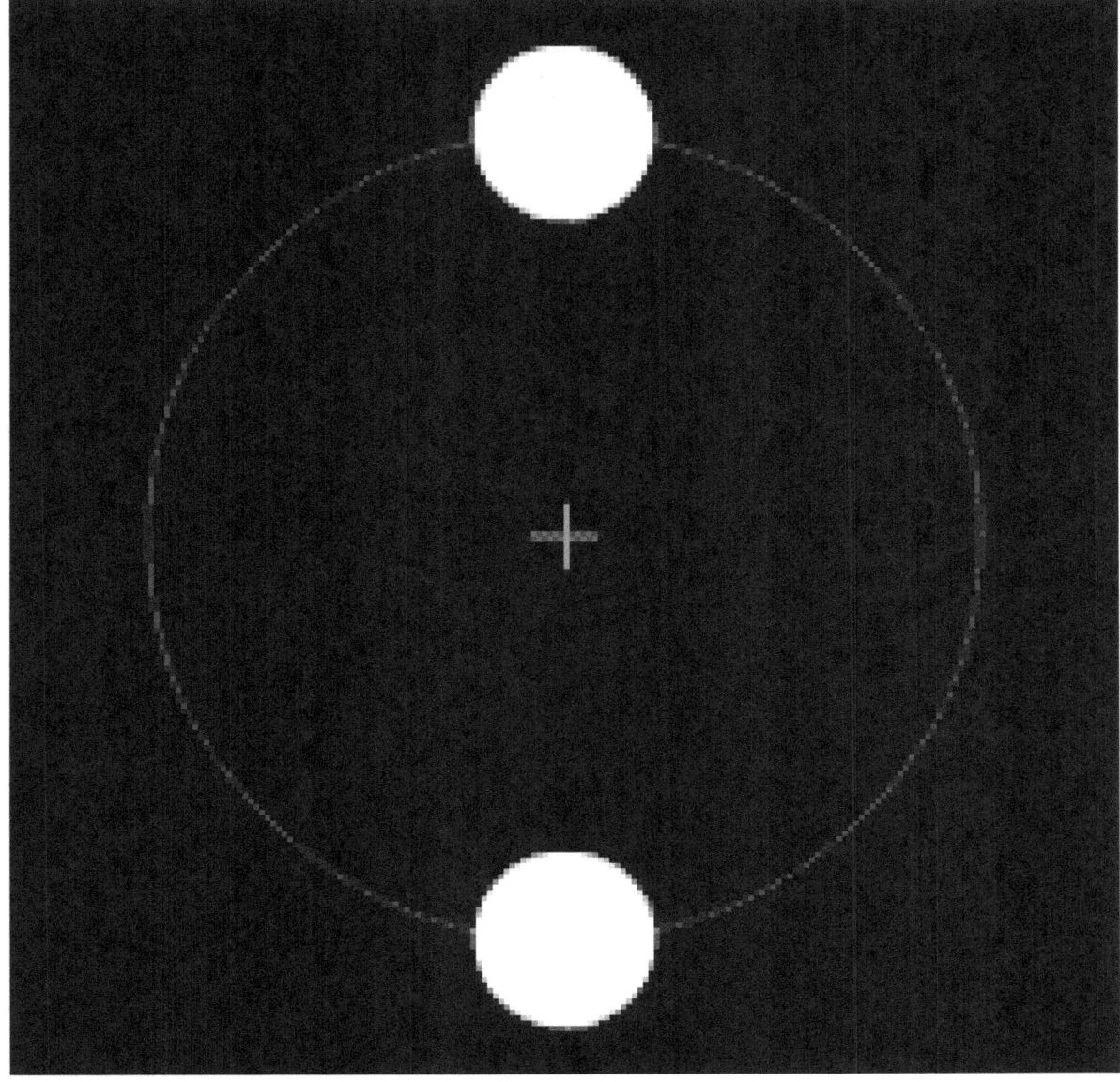

Two stars of similar mass are in circular orbits about their center of mass

The overall angular momentum is reduced however. This reduction corresponds to the angular momentum carried off by gravitational radiation. The rate of decrease of distance between the bodies versus time is given by:[26]

$$\frac{\mathrm{d}r}{\mathrm{d}t} = -\frac{64}{5}\,\frac{G^3}{c^5}\,\frac{(m_1 m_2)(m_1 + m_2)}{r^3}$$

where the variables are the same as in the previous equation.

The orbit decays at a rate proportional to the inverse third power of the radius. When the radius has shrunk to half its initial value, it is shrinking eight times faster than before. By Kepler's Third Law, the new rotation rate at this point will be faster by $\sqrt{8} = 2.828$, or nearly three times the previous orbital frequency. As the radius decreases, the power lost to gravitational radiation increases even more. As can be seen from the previous equation, power radiated varies as the inverse fifth power of the radius, or 32 times more in this case.

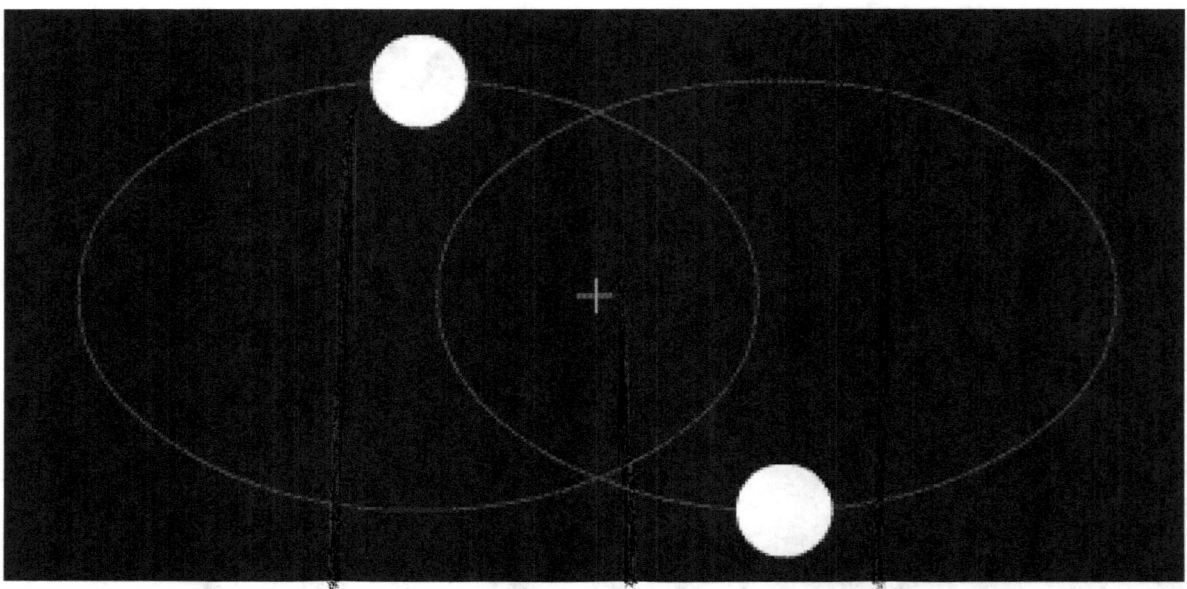

Two stars of similar mass are in highly elliptical orbits about their center of mass

If we use the previous values for the Sun and the Earth, we find that the Earth's orbit shrinks by 1.1×10^{-20} meter per second. This is 3.5×10^{-13} m per year, which is about 1/300 the diameter of a hydrogen atom. The effect of gravitational radiation on the size of the Earth's orbit is unnoticeable over the age of the universe. Actually, this effect is absolutely negligible compared to the increase of Earth's orbit due to losing mass of Sun via radiation (4.7×10^{-9} m/s).[27] This is not true for closer orbits.

A more practical example is the orbit of a Sun-like star around a heavy black hole. Our Milky Way is believed to have a 4 million $M\odot$ black hole at its center in Sagittarius A. Such supermassive black holes are being found in the center of almost all galaxies. For this example take a 2 million $M\odot$ black hole with a solar-mass star orbiting it at a radius of 1.89×10^{10} m (63 light-seconds). The mass of the black hole will be 4×10^{36} kg and its gravitational radius will be 6×10^{9} m. The orbital period will be 1,000 seconds, or a little under 17 minutes. The solar-mass star will draw closer to the black hole by 7.4 meters per second or 7.4 km per orbit. A collision will not be long in coming.

Assume that a pair of 1 $M\odot$ neutron stars are in circular orbits at a distance of 1.89×10^{8} m (189,000 km). This is a little less than 1/7 the diameter of the Sun or 0.63 light-seconds. Their orbital period would be 1,000 seconds. Substituting the new mass and radius in the above formula gives a rate of orbit decrease of 3.7×10^{-6} m/s or 3.7 mm per orbit. This is 116 meters per year and is not negligible over cosmic time scales.

Suppose instead that these two neutron stars were orbiting at a distance of 1.89×10^{6} m (1890 km). Their period would be 1 second and their orbital velocity would be about 1/50 of the speed of light. Their orbit would now shrink by 3.7 meters per orbit. A collision is imminent. A runaway loss of energy from the orbit results in an ever more rapid decrease in the distance between the stars. They will eventually merge to form a black hole and cease to radiate gravitational waves. This is referred to as the inspiral.

The above equation can not be applied directly for calculating the lifetime of the orbit, because the rate of change in radius depends on the radius itself, and is thus non-constant with time. The lifetime can be computed by integration of this equation (see next section).

1.3.3 Orbital lifetime limits from gravitational radiation

Orbital lifetime is one of the most important properties of gravitational radiation sources. It determines the average number of binary stars in the universe that are close enough to be detected. Short lifetime binaries are strong sources of gravitational radiation but are few in number. Long lifetime binaries are more plentiful but they are weak sources of gravitational waves. LIGO is most sensitive in the frequency band where two neutron stars are about to merge. This time frame is only a few seconds. It takes luck for the detector to see this blink in time out of a million year orbital lifetime. It is predicted that such a merger will only be seen once per decade or so.

The lifetime of an orbit is given by:[26]

$$t = \frac{5}{256} \frac{c^5}{G^3} \frac{r^4}{(m_1 m_2)(m_1 + m_2)}$$

where r is the initial distance between the orbiting bodies. This equation can be derived by integrating the previous equation for the rate of radius decrease. It predicts the time for the radius of the orbit to shrink to zero. As the orbital speed becomes a significant fraction of the speed of light, this equation becomes inaccurate. It is useful for inspirals until the last few milliseconds before the merger of the objects.

Substituting the values for the mass of the Sun and Earth as well as the orbital radius gives a very large lifetime of 3.44×10^{30} seconds or 1.09×10^{23} years (that is approximately 10^{13} times larger than the age of the universe). The actual figure would be slightly less than that. The Earth will break apart from tidal forces if it orbits closer than a few radii from the Sun. This would form a ring around the Sun and instantly stop the emission of gravitational waves.

If we use a 2 million $M\odot$ black hole with a solar mass star orbiting it at 1.89×10^{10} meters, we get a lifetime of 6.50×10^8 seconds or 20.7 years.

Assume that a pair of solar mass neutron stars with a diameter of 10 kilometers are in circular orbits at a distance of 1.89×10^8 m (189,000 km). Their lifetime is 1.30×10^{13} seconds or about 414,000 years. Their orbital period will be 1,000 seconds and it could be observed by LISA if they were not too far away. A far greater number of white dwarf binaries exist with orbital periods in this range. White dwarf binaries have masses on the order of our Sun and diameters on the order of our Earth. They cannot get much closer together than 10,000 km before they will merge and cease to radiate gravitational waves. This results in the creation of either a neutron star or a black hole. Until then, their gravitational radiation will be comparable to that of a neutron star binary. LISA is the only gravitational wave experiment that is likely to succeed in detecting such types of binaries.

If the orbit of a neutron star binary has decayed to 1.89×10^6m (1890 km), its remaining lifetime is 130,000 seconds or about 36 hours. The orbital frequency will vary from 1 revolution per second at the start and 918 revolutions per second when the orbit has shrunk to 20 km at merger. The gravitational radiation emitted will be at twice the orbital frequency. Just before merger, the inspiral can be observed by LIGO if the binary is close enough. LIGO has only a few minutes to observe this merger out of a total orbital lifetime that may have been billions of years. The chance of success with LIGO as initially constructed is quite low despite the large number of such mergers occurring in the universe, because the sensitivity of the instrument does not 'reach' out to enough systems to see events frequently. No mergers have been seen in the few years that initial LIGO has been in operation, and it is thought that a merger should be seen about once per several tens of years of observing time with initial LIGO.[28] The upgraded Advanced LIGO detector, with a ten times greater sensitivity, 'reaches' out 10 times further—encompassing a volume 1000 times greater, and seeing 1000 times as many candidate sources. Thus, the expectation is that detections will be made at the rate of tens per year.

1.3.4   Wave amplitudes from the Earth–Sun system

We can also think in terms of the amplitude of the wave from a system in circular orbits. Let θ be the angle between the perpendicular to the plane of the orbit and the line of sight of the observer. Suppose that an observer is outside the system at a distance R from its center of mass. If R is much greater than a wavelength, the two polarizations of the wave will be

$$h_+ = -\frac{1}{R} \frac{G^2}{c^4} \frac{2m_1 m_2}{r} (1 + \cos^2 \theta) \cos [2\omega(t - R)],$$

$$h_\times = -\frac{1}{R} \frac{G^2}{c^4} \frac{4m_1 m_2}{r} (\cos \theta) \sin [2\omega(t - R)].$$

Here, we use the constant angular velocity of a circular orbit in Newtonian physics:

$$\omega = \sqrt{G(m_1 + m_2)/r^3}.$$

For example, if the observer is in the x - y plane then $\theta = \pi/2$, and $\cos(\theta) = 0$, so the $h_\times$ polarization is always zero. We also see that the frequency of the wave given off is twice the rotation frequency. If we put in numbers for the Earth-Sun system, we find:

$$h_+ = -\frac{1}{R}\frac{G^2}{c^4}\frac{4m_1m_2}{r} = -\frac{1}{R}1.7 \times 10^{-10} \text{ m}.$$

In this case, the minimum distance to find waves is $R \approx 1$ light-year, so typical amplitudes will be $h \approx 10^{-26}$. That is, a ring of particles would stretch or squeeze by just one part in 10^{26}. This is well under the detectability limit of all conceivable detectors.

1.3.5 Radiation from other sources

Although the waves from the Earth-Sun system are minuscule, astronomers can point to other sources for which the radiation should be substantial. One important example is the Hulse-Taylor binary — a pair of stars, one of which is a pulsar.[29] The characteristics of their orbit can be deduced from the Doppler shifting of radio signals given off by the pulsar. Each of the stars are about 1.4 $M\odot$ and the size of their orbit is about 1/75 of the Earth-Sun orbit. This means the distance between the two stars is just a few times larger than the diameter of our own Sun. The combination of greater masses and smaller separation means that the energy given off by the Hulse-Taylor binary will be far greater than the energy given off by the Earth-Sun system — roughly 10^{22} times as much.

The information about the orbit can be used to predict just how much energy (and angular momentum) should be given off in the form of gravitational waves. As the energy is carried off, the stars should draw closer to each other. This effect is called an inspiral, and it can be observed in the pulsar's signals. The measurements on the Hulse-Taylor system have been carried out over more than 30 years. It has been shown that the gravitational radiation predicted by general relativity allows these observations to be matched within 0.2 percent. In 1993, Russell Hulse and Joe Taylor were awarded the Nobel Prize in Physics for this work, which was the first indirect evidence for gravitational waves. The orbital lifetime of this binary system before merger is a few hundred million years.[30]

Inspirals are very important sources of gravitational waves. Any time two compact objects (white dwarfs, neutron stars, or black holes) are in close orbits, they send out intense gravitational waves. As they spiral closer to each other, these waves become more intense. At some point they should become so intense that direct detection by their effect on objects on Earth or in space is possible. This direct detection is the goal of several large scale experiments.[31]

The only difficulty is that most systems like the Hulse-Taylor binary are so far away. The amplitude of waves given off by the Hulse-Taylor binary as seen on Earth would be roughly $h \approx 10^{-26}$. There are some sources, however, that astrophysicists expect to find with much larger amplitudes of $h \approx 10^{-20}$. At least eight other binary pulsars have been discovered.[32]

1.4 Astrophysics and gravitational waves

During the past century, astronomy has been revolutionized by the use of new methods for observing the universe. Astronomical observations were originally made using visible light. Galileo Galilei pioneered the use of telescopes to enhance these observations. However, visible light is only a small portion of the electromagnetic spectrum, and not all objects in the distant universe shine strongly in this particular band. More useful information may be found, for example, in radio wavelengths. Using radio telescopes, astronomers have found pulsars, quasars, and other extreme objects that push the limits of our understanding of physics. Observations in the microwave band have opened our eyes to the faint imprints of the Big Bang, a discovery Stephen Hawking called the "greatest discovery of the century, if not all time". Similar advances in observations using gamma rays, x-rays, ultraviolet light, and infrared light have also brought new insights to

Two-dimensional representation of gravitational waves generated by two neutron stars orbiting each other.

astronomy. As each of these regions of the spectrum has opened, new discoveries have been made that could not have been made otherwise. Astronomers hope that the same holds true of gravitational waves.

Gravitational waves have two important and unique properties. First, there is no need for any type of matter to be present nearby in order for the waves to be generated by a binary system of uncharged black holes, which would emit no electromagnetic radiation. Second, gravitational waves can pass through any intervening matter without being scattered significantly. Whereas light from distant stars may be blocked out by interstellar dust, for example, gravitational waves will pass through essentially unimpeded. These two features allow gravitational waves to carry information about astronomical phenomena never before observed by humans.

The sources of gravitational waves described above are in the low-frequency end of the gravitational-wave spectrum (10^{-7} to 10^5 Hz). An astrophysical source at the high-frequency end of the gravitational-wave spectrum (above 10^5 Hz and probably 10^{10} Hz) generates relic gravitational waves that are theorized to be faint imprints of the Big Bang like the cosmic microwave background (see gravitational wave background).[33] At these high frequencies it is potentially possible that the sources may be "man made"[23] that is, gravitational waves generated and detected in the laboratory.[34][35]

1.4.1   Energy, momentum, and angular momentum carried by gravitational waves

Waves familiar from other areas of physics such as water waves, sound waves, and electromagnetic waves are able to carry energy, momentum, and angular momentum. By carrying these away from a source, waves are able to rob that source of its energy as well as its linear and angular momentum. Gravitational waves perform the same function. Thus, for example, a binary system loses angular momentum as the two orbiting objects spiral towards each other—the angular momentum is radiated away by gravitational waves.

The waves can also carry off linear momentum, a possibility that has some interesting implications for astrophysics.[36] After two supermassive black holes coalesce, emission of linear momentum can produce a "kick" with amplitude as large as 4000 km/s. This is fast enough to eject the coalesced black hole completely from its host galaxy. Even if the kick is too small to eject the black hole completely, it can remove it temporarily from the nucleus of the galaxy, after which it will oscillate about the center, eventually coming to rest.[37] A kicked black hole can also carry a star cluster with it, forming a hyper-compact stellar system.[38] Or it may carry gas, allowing the recoiling black hole to appear temporarily as a "naked quasar". The quasar SDSS J092712.65+294344.0 is believed to contain a recoiling supermassive black hole.[39]

1.5 Detecting gravitational waves

1.5.1 Difficulties in detection

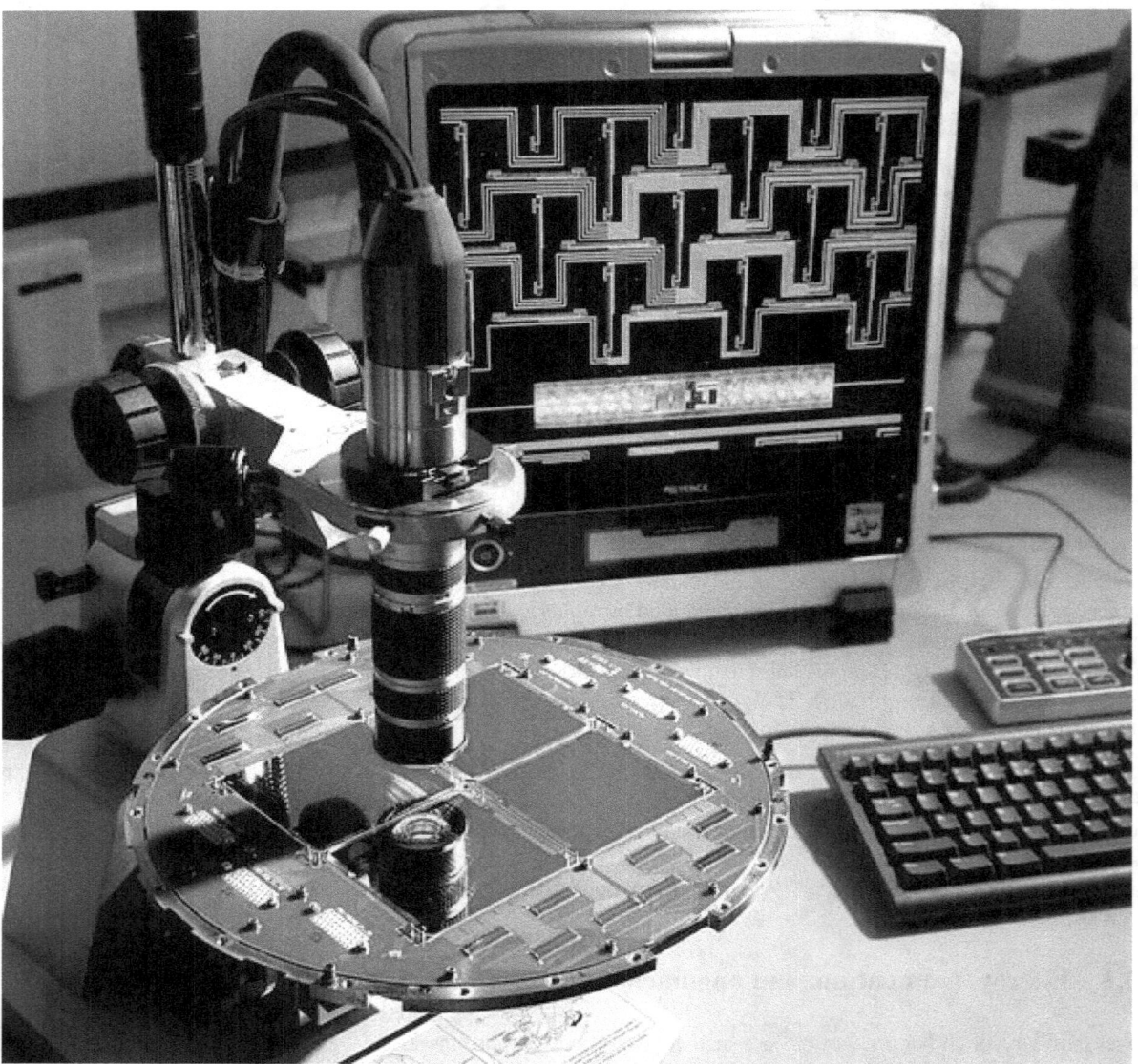

Evidence of gravitational waves in the infant universe may have been uncovered by the BICEP2 radio telescope. The microscopic examination of the focal plane of the BICEP2 detector is shown here.[7][8][9][10][40] In 2015, however, the BICEP2 findings were confirmed to be the result of dust.[18]

Gravitational waves are not easily detectable. This knowledge gap is primarily due to the massive presence of noise in the low frequencies where antennas currently operate. Gravitational waves are expected to have frequencies 10^{-16} Hz $<$ $f < 10^4$ Hz .[41]

1.5.2   Ground-based interferometers

Main article: Gravitational wave detector

Though the Hulse-Taylor observations were very important, they give only *indirect* evidence for gravitational waves. A more conclusive observation would be a *direct* measurement of the effect of a passing gravitational wave, which could also provide more information about the system that generated it. Any such direct detection is complicated by the extraordinarily small effect the waves would produce on a detector. The amplitude of a spherical wave will fall off as the inverse of the distance from the source (the $1/R$ term in the formulas for h above). Thus, even waves from extreme systems like merging binary black holes die out to very small amplitude by the time they reach the Earth. Astrophysicists expect that some gravitational waves passing the Earth may be as large as $h \approx 10^{-20}$, but generally no bigger.[42]

A simple device theorised to detect the expected wave motion is called a Weber bar — a large, solid bar of metal isolated from outside vibrations. This type of instrument was the first type of gravitational wave detector. Strains in space due to an incident gravitational wave excite the bar's resonant frequency and could thus be amplified to detectable levels. Conceivably, a nearby supernova might be strong enough to be seen without resonant amplification. With this instrument, Joseph Weber claimed to have detected daily signals of gravitational waves. His results, however, were contested in 1974 by physicists Richard Garwin and David Douglass. Modern forms of the Weber bar are still operated, cryogenically cooled, with superconducting quantum interference devices to detect vibration. Weber bars are not sensitive enough to detect anything but extremely powerful gravitational waves.[43]

MiniGRAIL is a spherical gravitational wave antenna using this principle. It is based at Leiden University, consisting of an exactly machined 1150 kg sphere cryogenically cooled to 20 mK.[44] The spherical configuration allows for equal sensitivity in all directions, and is somewhat experimentally simpler than larger linear devices requiring high vacuum. Events are detected by measuring deformation of the detector sphere. MiniGRAIL is highly sensitive in the 2–4 kHz range, suitable for detecting gravitational waves from rotating neutron star instabilities or small black hole mergers.[45]

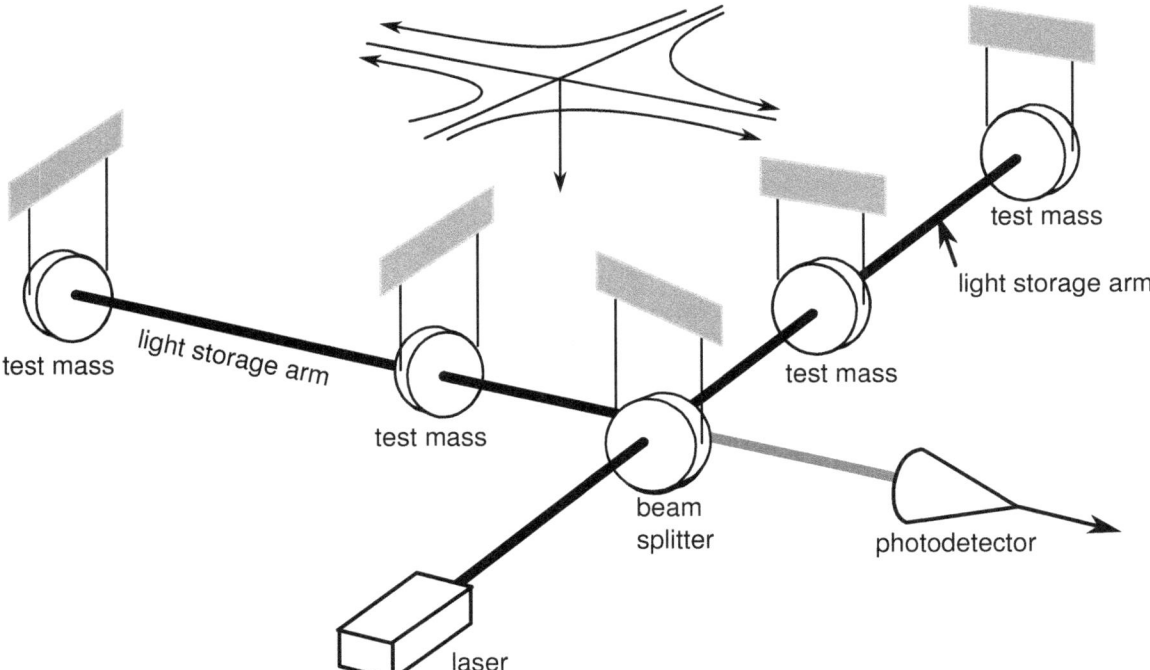

A schematic diagram of a laser interferometer

A more sensitive class of detector uses laser interferometry to measure gravitational-wave induced motion between separated 'free' masses.[46] This allows the masses to be separated by large distances (increasing the signal size); a further advantage is that it is sensitive to a wide range of frequencies (not just those near a resonance as is the case for Weber bars). Ground-based interferometers are now operational. Currently, the most sensitive is LIGO — the Laser Interferometer Gravitational Wave Observatory. LIGO has three detectors: one in Livingston, Louisiana; the other two (in the same vacuum tubes) at the Hanford site in Richland, Washington. Each consists of two light storage arms that are 2 to 4 kilometers in length. These are at 90 degree angles to each other, with the light passing through 1m diameter vacuum tubes running the entire 4 kilometers. A passing gravitational wave will slightly stretch one arm as it shortens the other. This is precisely the motion to which an interferometer is most sensitive.

Even with such long arms, the strongest gravitational waves will only change the distance between the ends of the arms by at most roughly 10^{-18} meters. LIGO should be able to detect gravitational waves as small as $h \sim 5 \times 10^{-20}$. Upgrades to LIGO and other detectors such as Virgo, GEO 600, and TAMA 300 should increase the sensitivity still further; the next generation of instruments (Advanced LIGO and Advanced Virgo) will be more than ten times more sensitive. Another highly sensitive interferometer, KAGRA, is under construction in the Kamiokande mine in Japan. A key point is that a tenfold increase in sensitivity (radius of 'reach') increases the volume of space accessible to the instrument by one thousand times. This increases the rate at which detectable signals should be seen from one per tens of years of observation, to tens per year.[28]

Interferometric detectors are limited at high frequencies by shot noise, which occurs because the lasers produce photons randomly; one analogy is to rainfall—the rate of rainfall, like the laser intensity, is measurable, but the raindrops, like photons, fall at random times, causing fluctuations around the average value. This leads to noise at the output of the detector, much like radio static. In addition, for sufficiently high laser power, the random momentum transferred to the test masses by the laser photons shakes the mirrors, masking signals at low frequencies. Thermal noise (e.g., Brownian motion) is another limit to sensitivity. In addition to these 'stationary' (constant) noise sources, all ground-based detectors are also limited at low frequencies by seismic noise and other forms of environmental vibration, and other 'non-stationary' noise sources; creaks in mechanical structures, lightning or other large electrical disturbances, etc. may also create noise masking an event or may even imitate an event. All these must be taken into account and excluded by analysis before a detection may be considered a true gravitational wave event.

Space-based interferometers, such as LISA and DECIGO, are also being developed. LISA's design calls for three test masses forming an equilateral triangle, with lasers from each spacecraft to each other spacecraft forming two independent interferometers. LISA is planned to occupy a solar orbit trailing the Earth, with each arm of the triangle being five million kilometers. This puts the detector in an excellent vacuum far from Earth-based sources of noise, though it will still be susceptible to shot noise, as well as artifacts caused by cosmic rays and solar wind.

There are currently two detectors focusing on detection at the higher end of the gravitational wave spectrum (10^{-7} to 10^5 Hz): one at University of Birmingham, England, and the other at INFN Genoa, Italy. A third is under development at Chongqing University, China. The Birmingham detector measures changes in the polarization state of a microwave beam circulating in a closed loop about one meter across. Two have been fabricated and they are currently expected to be sensitive to periodic spacetime strains of $h \sim 2 \times 10^{-13}/\sqrt{\text{Hz}}$, given as an amplitude spectral density. The INFN Genoa detector is a resonant antenna consisting of two coupled spherical superconducting harmonic oscillators a few centimeters in diameter. The oscillators are designed to have (when uncoupled) almost equal resonant frequencies. The system is currently expected to have a sensitivity to periodic spacetime strains of $h \sim 2 \times 10^{-17}/\sqrt{\text{Hz}}$, with an expectation to reach a sensitivity of $h \sim 2 \times 10^{-20}/\sqrt{\text{Hz}}$. The Chongqing University detector is planned to detect relic high-frequency gravitational waves with the predicted typical parameters $?_g \sim 10^{10}$ Hz (10 GHz) and $h \sim 10^{-30}-10^{-31}$.

1.5.3 Using pulsar timing arrays

Pulsars are rapidly rotating stars. A pulsar emits beams of radio waves that, like lighthouse beams, sweep through the sky as the pulsar rotates. The signal from a pulsar can be detected by radio telescopes as a series of regularly spaced pulses, essentially like the ticks of a clock. Gravitational waves affect the time it takes the pulses to travel from the pulsar to a telescope on Earth. A pulsar timing array uses millisecond pulsars to seek out perturbations due to gravitational waves in measurements of pulse arrival times at a telescope, in other words, to look for deviations in the clock ticks. In particular, pulsar timing arrays can search for a distinct pattern of correlation and anti-correlation between the signals over an array of different pulsars (resulting in the name "pulsar timing array"). Although pulsar pulses travel through space for hundreds or thousands of years to reach us, pulsar timing arrays are sensitive to perturbations in their travel time of much less than a millionth of a second.

Globally there are three active pulsar timing array projects. The North American Nanohertz Gravitational Wave Observatory uses data collected by the Arecibo Radio Telescope and Green Bank Telescope. The Parkes Pulsar Timing Array at the Parkes radio-telescope has been collecting data since March 2005. The European Pulsar Timing Array uses data from the four largest telescopes in Europe: the Lovell Telescope, the Westerbork Synthesis Radio Telescope, the Effelsberg Telescope and the Nancay Radio Telescope. (Upon completion the Sardinia Radio Telescope will be added to the EPTA also.) These three projects have begun collaborating under the title of the International Pulsar Timing Array project.

1.5.4   Einstein@Home

Main article: Einstein@Home

In some sense, the easiest signals to detect should be constant sources. Supernovae and neutron star or black hole mergers should have larger amplitudes and be more interesting, but the waves generated will be more complicated. The waves given off by a spinning, aspherical neutron star would be "monochromatic"—like a pure tone in acoustics. It would not change very much in amplitude or frequency.

The Einstein@Home project is a distributed computing project similar to SETI@home intended to detect this type of simple gravitational wave. By taking data from LIGO and GEO, and sending it out in little pieces to thousands of volunteers for parallel analysis on their home computers, Einstein@Home can sift through the data far more quickly than would be possible otherwise.[47]

1.5.5   Primordial gravitational waves

Main article: Primordial gravitational wave

Primordial gravitational waves are gravitational waves observed in the cosmic microwave background. They were allegedly detected by the BICEP2 instrument, an announcement made on 17 March 2014, which was withdrawn on 30 January 2015 ("the signal can be entirely attributed to dust in the Milky Way"[18]).

1.6   Mathematics

Einstein's equations form the fundamental law of general relativity. The curvature of spacetime can be expressed mathematically using the metric tensor — denoted $g_{\mu\nu}$. The metric holds information regarding how distances are measured in the space under consideration. Because the propagation of gravitational waves through space and time change distances, we will need to use this to find the solution to the wave equation.

Spacetime curvature is also expressed with respect to a covariant derivative, ∇, in the form of the Einstein tensor, $G_{\mu\nu}$. This curvature is related to the stress–energy tensor, $T_{\mu\nu}$, by the key equation

$$G_{\mu\nu} = \frac{8\pi G_N}{c^4} T_{\mu\nu},$$

where G_N is Newton's gravitational constant, and c is the speed of light. We assume geometrized units, so $G_N = 1 = c$.

With some simple assumptions, Einstein's equations can be rewritten to show explicitly that they are wave equations. To begin with, we adopt some coordinate system, like (t, r, θ, ϕ). We define the "flat-space metric" $\eta_{\mu\nu}$ to be the quantity that — in this coordinate system — has the components we would expect for the flat space metric. For example, in these spherical coordinates, we have

$$\eta_{\mu\nu} = \begin{bmatrix} -1 & 0 & 0 & 0 \\ 0 & 1 & 0 & 0 \\ 0 & 0 & r^2 & 0 \\ 0 & 0 & 0 & r^2 \sin^2 \theta \end{bmatrix}.$$

This flat-space metric has no physical significance; it is a purely mathematical device necessary for the analysis. Tensor indices are raised and lowered using this "flat-space metric".

Now, we can also think of the physical metric $g_{\mu\nu}$ as a matrix, and find its determinant, $\det g$. Finally, we define a quantity

$$\bar{h}^{\alpha\beta} \equiv \eta^{\alpha\beta} - \sqrt{|\det g|} g^{\alpha\beta}$$

This is the crucial field, which will represent the radiation. It is possible (at least in an asymptotically flat spacetime) to choose the coordinates in such a way that this quantity satisfies the "de Donder" gauge conditions (conditions on the coordinates):

$$\nabla_\beta \bar{h}^{\alpha\beta} = 0,$$

where ∇ represents the flat-space derivative operator. These equations say that the divergence of the field is zero. The linear Einstein equations can now be written[48] as

$$\Box \bar{h}^{\alpha\beta} = -16\pi \tau^{\alpha\beta}$$

where $\Box = -\partial_t^2 + \Delta$ represents the flat-space d'Alembertian operator, and $\tau^{\alpha\beta}$ represents the stress–energy tensor plus quadratic terms involving $\bar{h}^{\alpha\beta}$. This is just a wave equation for the field with a source, despite the fact that the source involves terms quadratic in the field itself. That is, it can be shown that solutions to this equation are waves traveling with velocity 1 in these coordinates.

1.6.1 Linear approximation

The equations above are valid everywhere — near a black hole, for instance. However, because of the complicated source term, the solution is generally too difficult to find analytically. We can often assume that space is nearly flat, so the metric is nearly equal to the $\eta^{\alpha\beta}$ tensor. In this case, we can neglect terms quadratic in $\bar{h}^{\alpha\beta}$, which means that the $\tau^{\alpha\beta}$ field reduces to the usual stress–energy tensor $T^{\alpha\beta}$. That is, Einstein's equations become

$$\Box \bar{h}^{\alpha\beta} = -16\pi T^{\alpha\beta}$$

If we are interested in the field far from a source, however, we can treat the source as a point source; everywhere else, the stress–energy tensor would be zero, so

$$\Box \bar{h}^{\alpha\beta} = 0$$

Now, this is the usual homogeneous wave equation — one for each component of $\bar{h}^{\alpha\beta}$. Solutions to this equation are well known. For a wave moving away from a point source, the radiated part (meaning the part that dies off as $1/r$ far from the source) can always be written in the form $A(t - r, \theta, \phi)/r$, where A is just some function. It can be shown[49] that — to a linear approximation — it is always possible to make the field traceless. Now, if we further assume that the source is positioned at $r = 0$, the general solution to the wave equation in spherical coordinates is

$$\bar{h}^{\alpha\beta} = \frac{1}{r} \begin{bmatrix} 0 & 0 & 0 & 0 \\ 0 & 0 & 0 & 0 \\ 0 & 0 & A_+(t-r,\theta,\phi) & A_\times(t-r,\theta,\phi) \\ 0 & 0 & A_\times(t-r,\theta,\phi) & -A_+(t-r,\theta,\phi) \end{bmatrix}$$

$$\equiv \begin{bmatrix} 0 & 0 & 0 & 0 \\ 0 & 0 & 0 & 0 \\ 0 & 0 & h_+(t-r,r,\theta,\phi) & h_\times(t-r,r,\theta,\phi) \\ 0 & 0 & h_\times(t-r,r,\theta,\phi) & -h_+(t-r,r,\theta,\phi) \end{bmatrix}$$

where we now see the origin of the two polarizations.

1.6.2   Relation to the source

If we know the details of a source — for instance, the parameters of the orbit of a binary — we can relate the source's motion to the gravitational radiation observed far away. With the relation

$$\Box \bar{h}^{\alpha\beta} = -16\pi \tau^{\alpha\beta}$$

we can write the solution in terms of the tensorial Green's function for the d'Alembertian operator:[48]

$$\bar{h}^{\alpha\beta}(t,\vec{x}) = -16\pi \int G^{\alpha\beta}_{\gamma\delta}(t,\vec{x};t',\vec{x}')\, \tau^{\gamma\delta}(t',\vec{x}')\, \mathrm{d}t'\, \mathrm{d}^3x'$$

Though it is possible to expand the Green's function in tensor spherical harmonics, it is easier to simply use the form

$$G^{\alpha\beta}_{\gamma\delta}(t,\vec{x};t',\vec{x}') = \frac{1}{4\pi} \delta^\alpha_\gamma\, \delta^\beta_\delta\, \frac{\delta(t \pm |\vec{x} - \vec{x}'| - t')}{|\vec{x} - \vec{x}'|}$$

where the positive and negative signs correspond to ingoing and outgoing solutions, respectively. Generally, we are interested in the outgoing solutions, so

$$\bar{h}^{\alpha\beta}(t,\vec{x}) = -4 \int \frac{\tau^{\alpha\beta}(t - |\vec{x} - \vec{x}'|, \vec{x}')}{|\vec{x} - \vec{x}'|}\, \mathrm{d}^3x'$$

If the source is confined to a small region very far away, to an excellent approximation we have:

$$\bar{h}^{\alpha\beta}(t,\vec{x}) \approx -\frac{4}{r} \int \tau^{\alpha\beta}(t - r, \vec{x}')\, \mathrm{d}^3x'$$

where $r = |\vec{x}|$.

Now, because we will eventually only be interested in the spatial components of this equation (time components can be set to zero with a coordinate transformation), and we are integrating this quantity — presumably over a region of which there is no boundary — we can put this in a different form. Ignoring divergences with the help of Stokes' theorem and an empty boundary, we can see that

$$\int \tau^{ij}(t - r, \vec{x}')\, \mathrm{d}^3x' = \int x'^i x'^j \nabla_k \nabla_l \tau^{kl}(t - r, \vec{x}')\, \mathrm{d}^3x'$$

Inserting this into the above equation, we arrive at

$$\bar{h}^{ij}(t, \vec{x}) \approx -\frac{4}{r} \int x'^i x'^j \nabla_k \nabla_l \tau^{kl}(t - r, \vec{x}') \, \mathrm{d}^3 x'$$

Finally, because we have chosen to work in coordinates for which $\nabla_\beta \bar{h}^{\alpha\beta} = 0$, we know that $\nabla_\beta \tau^{\alpha\beta} = 0$. With a few simple manipulations, we can use this to prove that

$$\nabla_0 \nabla_0 \tau^{00} = \nabla_j \nabla_k \tau^{jk}$$

With this relation, the expression for the radiated field is

$$\bar{h}^{ij}(t, \vec{x}) \approx -\frac{4}{r} \frac{\mathrm{d}^2}{\mathrm{d}t^2} \int x'^i x'^j \tau^{00}(t - r, \vec{x}') \, \mathrm{d}^3 x'$$

In the linear case, $\tau^{00} = \rho$, the density of mass-energy.

To a very good approximation, the density of a simple binary can be described by a pair of delta-functions, which eliminates the integral. Explicitly, if the masses of the two objects are M_1 and M_2, and the positions are $\vec{x}_1$ and $\vec{x}_2$, then

$$\rho(t - r, \vec{x}') = M_1 \delta^3(\vec{x}' - \vec{x}_1(t - r)) + M_2 \delta^3(\vec{x}' - \vec{x}_2(t - r))$$

We can use this expression to do the integral above:

$$\bar{h}^{ij}(t, \vec{x}) \approx -\frac{4}{r} \frac{\mathrm{d}^2}{\mathrm{d}t^2} \left\{ M_1 x_1^i(t - r) x_1^j(t - r) + M_2 x_2^i(t - r) x_2^j(t - r) \right\}$$

Using mass-centered coordinates, and assuming a circular binary, this is

$$\bar{h}^{ij}(t, \vec{x}) \approx -\frac{4}{r} \frac{M_1 M_2}{R} n^i(t - r) n^j(t - r)$$

where $\vec{n} = \vec{x}_1 / |\vec{x}_1|$. Plugging in the known values of $\vec{x}_1(t - r)$, we obtain the expressions given above for the radiation from a simple binary.

1.7 See also

- Gravitational wave background

- Cosmic gravitational wave background

- Big Bang Observer (BBO), proposed successor to LISA

- DECIGO "Deci-hertz Interferometer Gravitational wave Observatory", the planned laser interferometric detector in space

- Gravitational field

- Gravitomagnetism

- Graviton

- Gravitational wave astronomy

- Hawking radiation, for gravitationally induced electromagnetic radiation from black holes

- HM Cancri

- LIGO, VIRGO, GEO 600, and TAMA 300 — Gravitational wave detectors

- Linearised Einstein field equations

- LISA the proposed Laser Interferometer Space Antenna

- Peres metric

- pp-wave spacetime, for an important class of exact solutions modelling gravitational radiation

- Spin-flip, a consequence of gravitational wave emission from binary supermassive black holes

- Sticky bead argument, for a physical way to see that gravitational radiation should carry energy

- Tidal force

1.8   References

[1] Einstein, A (June 1916). "Näherungsweise Integration der Feldgleichungen der Gravitation". *Sitzungsberichte der Königlich Preussischen Akademie der Wissenschaften Berlin*. part 1: 688–696.

[2] Einstein, A (1918). "Über Gravitationswellen". *Sitzungsberichte der Königlich Preussischen Akademie der Wissenschaften Berlin*. part 1: 154–167.

[3] Finley, Dave. "Einstein's gravity theory passes toughest test yet: Bizarre binary star system pushes study of relativity to new limits.". Phys.Org.

[4] The Detection of Gravitational Waves using LIGO, B. Barish

[5] "First Second of the Big Bang". *How The Universe Works 3*. 2014. Discovery Science.

[6] http://www.theguardian.com/science/2014/jun/04/gravitational-wave-discovery-dust-big-bang-inflation

[7] Staff (17 March 2014). "BICEP2 2014 Results Release". *National Science Foundation*. Retrieved 18 March 2014.

[8] "First Direct Evidence of Cosmic Inflation". *http://www.cfa.harvard.edu*. Harvard-Smithsonian Center for Astrophysics. 17 March 2014. Retrieved 17 March 2014.

[9] Clavin, Whitney (17 March 2014). "NASA Technology Views Birth of the Universe". *NASA*. Retrieved 17 March 2014.

[10] Overbye, Dennis (17 March 2014). "Detection of Waves in Space Buttresses Landmark Theory of Big Bang". *New York Times*. Retrieved 17 March 2014.

[11] Overbye, Dennis (19 June 2014). "Astronomers Hedge on Big Bang Detection Claim". *New York Times*. Retrieved 20 June 2014.

[12] Amos, Jonathan (19 June 2014). "Cosmic inflation: Confidence lowered for Big Bang signal". *BBC News*. Retrieved 20 June 2014.

[13] Ade, P.A.R. et al. (BICEP2 Collaboration) (19 June 2014). "Detection of B-Mode Polarization at Degree Angular Scales by BICEP2" (PDF). *Physical Review Letters* **112**: 241101. arXiv:1403.3985. Bibcode:2014PhRvL.112x1101A. doi:10.1103/PhysRevLett.112.241101.PMID24996078.Retrieved20June2014.

[14] Planck Collaboration Team (19 September 2014). "Planck intermediate results. XXX. The angular power spectrum of polarized dust emission at intermediate and high Galactic latitudes". *ArXiv*. arXiv:1409.5738. Bibcode:2014arXiv1409.5738P. Retrieved 22 September 2014.

[15] Overbye, Dennis (22 September 2014). "Study Confirms Criticism of Big Bang Finding". *New York Times.* Retrieved 22 September 2014.

[16] Clavin, Whitney (30 January 2015). "Gravitational Waves from Early Universe Remain Elusive". *NASA.* Retrieved 30 January 2015.

[17] Overbye, Dennis (30 January 2015). "Speck of Interstellar Dust Obscures Glimpse of Big Bang". *New York Times.* Retrieved 31 January 2015.

[18] Cowen, Ron (2015-01-30). "Gravitational waves discovery now officially dead". *nature.* doi:10.1038/nature.2015.16830.

[19] LIGO Scientific Collaboration; Virgo Collaboration (2012). "Search for Gravitational Waves from Low Mass Compact Binary Coalescence in LIGO's Sixth Science Run and Virgo's Science Runs 2 and 3". *Physical Review D* **85**: 082002. arXiv:1111.7314. Bibcode:2012PhRvD..85h2002A. doi:10.1103/PhysRevD.85.082002.

[20] LIGO Scientific Collaboration; Virgo Collaboration (2012). "All-sky search for gravitational-wave bursts in the second joint LIGO-Virgo run". *Physical Review D* **85**: 122007. arXiv:1202.2788. Bibcode:2012PhRvD..85l2007A. doi:10.1103/PhysRev D.85.122007.

[21] LIGO Scientific Collaboration; Virgo Collaboration (2013). "Search for gravitational waves from binary black hole inspiral, merger, and ringdown in LIGO-Virgo data from 2009-2010". *Physical Review D* **87**: 022002. arXiv:1209.6533. Bibcode:2013 PhRvD..87b2002A.doi:10.1103/PhysRevD.87.022002.

[22] Krauss, LM; Dodelson, S; Meyer, S (2010). "Primordial Gravitational Waves and Cosmology". *Science* **328** (5981): 989–992. arXiv:1004.2504. Bibcode:2010Sci...328..989K. doi:10.1126/science.1179541. PMID 20489015.

[23] Hawking, S. W. and Israel, W., *General Relativity: An Einstein Centenary Survey*, Cambridge University Press, Cambridge, 1979, 98.

[24] Landau, L. D. and Lifshitz, E. M., *The Classical Theory of Fields.* Fourth Revised English Edition, Pergamon Press., 1975, 356–357.

[25] Einstein, A (1918). "Über Gravitationswellen". *Sitzungsberichte, Preussische Akademie der Wissenschaften* **154**.

[26] Gravitational Radiation

[27] Beatty, Kelly. "Why is the Earth moving away from the sun?". *New Scientist.*

[28] LIGO Scientific Collaboration; Virgo Collaboration (2010). "Predictions for the rates of compact binary coalescences observable by ground-based gravitational-wave detectors". *Classical and Quantum Gravity* **27**: 17300. arXiv:1003.2480. Bibcode:2010 CQGra..27q3001A.doi:10.1088/0264-9381/27/17/173001.

[29] Relativistic Binary Pulsar B1913+16: Thirty Years of Observations and Analysis

[30] The discovery of the first binary pulsar

[31] Crashing Black Holes

[32] Binary and Millisecond Pulsars

[33] L. P. Grishchuk (1976), "Primordial Gravitons and the Possibility of Their Observation", Sov. Phys. JETP Lett. 23, p. 293.

[34] Braginsky, V. B., Rudenko and Valentin, N. Section 7: "Generation of gravitational waves in the laboratory", *Physics Report* (Review section of *Physics Letters*), 46, No. 5. 165–200, (1978).

[35] Li, Fangyu, Baker, R. M L, Jr., and Woods, R. C., "Piezoelectric-Crystal-Resonator High-Frequency Gravitational Wave Generation and Synchro-Resonance Detection", in the proceedings of *Space Technology and Applications International Forum (STAIF-2006)*, edited by M.S. El-Genk, American Institute of Physics Conference Proceedings, Melville NY 813: 2006.

[36] Merritt, D.; et al. (May 2004). "Consequences of Gravitational Wave Recoil". *The Astrophysical Journal Letters* **607** (1): L9–L12. arXiv:astro-ph/0402057. Bibcode:2004ApJ...607L...9M. doi:10.1086/421551.

[37] Gualandris, A.; Merritt, D.; et al. (May 2008). "Ejection of Supermassive Black Holes from Galaxy Cores". *The Astrophysical Journal* **678** (2): 780–797. arXiv:0708.0771. Bibcode:2008ApJ...678..780G. doi:10.1086/586877.

[38] Merritt, D.; Schnittman, J. D.; Komossa, S. (2009). "Hypercompact Stellar Systems Around Recoiling Supermassive Black Holes". *The Astrophysical Journal* **699** (2): 1690–1710. arXiv:0809.5046. Bibcode:2009ApJ...699.1690M. doi:10.1088/0004-637X/699/2/1690.

[39] Komossa, S.; Zhou, H.; Lu, H. (May 2008). "A Recoiling Supermassive Black Hole in the Quasar SDSS J092712.65+294344.0?". *The Astrophysical Journal* **678** (2): L81–L84. arXiv:0804.4585. Bibcode:2008ApJ...678L..81K. doi:10.1086/588656

[40] Overbye, Dennis (24 March 2014). "Ripples From the Big Bang". *New York Times*. Retrieved 24 March 2014.

[41] Thorne, Kip S. (1995). "Gravitational Waves". *Cornell University Library*.

[42] David G. Blair (Ed.) (1991). *The detection of gravitational waves*. Cambridge University Press.

[43] For a review of early experiments using Weber bars, see Levine, J. (April 2004). "Early Gravity-Wave Detection Experiments, 1960–1975". *Physics in Perspective (Birkhäuser Basel)* **6** (1): 42–75. Bibcode:2004PhP.....6...42L. doi:10.1007/s00016-003-0179-6.

[44] Gravitational Radiation Antenna In Leiden

[45] de Waard, Arlette; Luciano Gottardi; Giorgio Frossati (July 2000). *Spherical Gravitational Wave Detectors: cooling and quality factor of a small CuAl6% sphere* (PDF). Marcel Grossmann meeting on General Relativity. Rome, Italy: World Scientific Publishing Co. Pte. Ltd. (published December 2002). pp. 1899–1901. Bibcode:2002nmgm.meet.1899D. doi:10.1142/9789812777386_0420.ISBN9789812777386.

[46] The idea of using laser interferometry for gravitational wave detection was first mentioned by Gerstenstein and Pustovoit 1963 Sov. Phys.–JETP 16 433. Weber mentioned it in an unpublished laboratory notebook. Rainer Weiss first described in detail a practical solution with an analysis of realistic limitations to the technique in R. Weiss (1972). "Electromagetically Coupled Broadband Gravitational Antenna". Quarterly Progress Report, Research Laboratory of Electronics, MIT 105: 54.

[47] Einstein@Home

[48] Thorne, Kip (April 1980). "Multipole expansions of gravitational radiation". *Reviews of Modern Physics* **52** (2): 299–339. Bibcode:1980RvMP...52..299T. doi:10.1103/RevModPhys.52.299.

[49] C. W. Misner, K. S. Thorne, and J. A. Wheeler (1973). *Gravitation*. W. H. Freeman and Co.

1.9   Further reading

- Chakrabarty, Indrajit, "Gravitational Waves: An Introduction". arXiv:physics/9908041 v1, Aug 21, 1999.

- Landau, L. D. and Lifshitz, E. M., The Classical Theory of Fields (Pergamon Press),(1987).

- Will, Clifford M., *The Confrontation between General Relativity and Experiment*. Living Rev. Relativity 9 (2006) 3.

- Peter Saulson, "Fundamentals of Interferometric Gravitational Wave Detectors", World Scientific, 1994.

- J. Bicak, W.N. Rudienko, "Gravitacionnyje wolny w OTO i probliema ich obnarużenija", Izdatielstwo Moskovskovo Universitieta, 1987.

- A. Kułak, "Electromagnetic Detectors of Gravitational Radiation", PhD Thesis, Cracow 1980 (In Polish).

- P. Tatrocki, "On intuitive description of graviton detector", www.philica.com .

- P. Tatrocki, "Can the LIGO, VIRGO, GEO600, AIGO, TAMA, LISA detectors really detect?", www.philica.com .

1.10 Bibliography

- Berry, Michael, *Principles of Cosmology and Gravitation* (Adam Hilger, Philadelphia, 1989). ISBN 0-85274-037-9

- Collins, Harry, *Gravity's Shadow: The Search for Gravitational Waves*, University of Chicago Press, 2004.

- P. J. E. Peebles, *Principles of Physical Cosmology* (Princeton University Press, Princeton, 1993). ISBN 0-691-01933-9.

- Wheeler, John Archibald and Ciufolini, Ignazio, *Gravitation and Inertia* (Princeton University Press, Princeton, 1995). ISBN 0-691-03323-4.

- Woolf, Harry, ed., *Some Strangeness in the Proportion* (Addison–Wesley, Reading, Massachusetts, 1980). ISBN 0-201-09924-1.

1.11 External links

- Gravitational waves at *Encyclopædia Britannica*

-

- Gravitational Waves on *In Our Time* at the BBC. (listen now)

- The LISA Brownbag – Selection of the most significant e-prints related to LISA science

- Astroparticle.org. To know everything about astroparticle physics, including gravitational waves

- Caltech's Physics 237-2002 Gravitational Waves by Kip Thorne **Video plus notes:** Graduate level but does not assume knowledge of General Relativity, Tensor Analysis, or Differential Geometry; Part 1: Theory (10 lectures), Part 2: Detection (9 lectures)

- www.astronomycast.com January 14, 2008 Episode 71: Gravitational Waves

- Laser Interferometer Gravitational Wave Observatory. LIGO Laboratory, operated by the California Institute of Technology and the Massachusetts Institute of Technology

- The LIGO Scientific Collaboration

- Einstein's Messengers – The LIGO Movie by NSF

- Home page for Einstein@Home project, a distributed computing project processing raw data from LIGO Laboratory, searching for gravitational waves

- The National Center for Supercomputing Applications – a numerical relativity group

- Caltech Relativity Tutorial – A basic introduction to gravitational waves, and astrophysical systems giving off gravitational waves

- Resource Letter GrW-1: Gravitational waves – a list of books, journals and web resources compiled by Joan Centrella for research into gravitational waves

- Mathematical and Physical Perspectives on Gravitational Radiation – written by B F Schutz of the Max Planck Institute explaining the significance and background of some key concepts in gravitational radiation

- Binary BH Merger – estimating the radiated power and merger time of a BH binary using dimensional analysis

Chapter 2

Gravitational field

In physics, a **gravitational field** is a model used to explain the influence that a massive body extends into the space around itself, producing a force on another massive body.[1] Thus, a gravitational field is used to explain gravitational phenomena, and is measured in newtons per kilogram (N/kg). In its original concept, gravity was a force between point masses. Following Newton, Laplace attempted to model gravity as some kind of radiation field or fluid, and since the 19th century explanations for gravity have usually been taught in terms of a field model, rather than a point attraction.

In a field model, rather than two particles attracting each other, the particles distort spacetime via their mass, and this distortion is what is perceived and measured as a "force". In such a model one states that matter moves in certain ways in response to the curvature of spacetime,[2] and that there is either *no gravitational force*,[3] or that gravity is a fictitious force.[4]

2.1   Classical mechanics

In classical mechanics as in physics, a gravitational field is a physical quantity.[5] A gravitational field can be defined using Newton's law of universal gravitation. Determined in this way, the gravitational field **g** around a single particle of mass M is a vector field consisting at every point of a vector pointing directly towards the particle. The magnitude of the field at every point is calculated applying the universal law, and represents the force per unit mass on any object at that point in space. Because the force field is conservative, there is a scalar potential energy per unit mass, Φ, at each point in space associated with the force fields; this is called gravitational potential.[6] The gravitational field equation is[7]

$$\mathbf{g} = \frac{\mathbf{F}}{m} = -\frac{\mathrm{d}^2\mathbf{R}}{\mathrm{d}t^2} = -GM\frac{\hat{\mathbf{R}}}{|\mathbf{R}|^2} = -\nabla\Phi,$$

where **F** is the gravitational force, m is the mass of the test particle, **R** is the position of the test particle, $\hat{\mathbf{R}}$ is a unit vector in the direction of **R**, t is time, G is the gravitational constant, and ∇ is the del operator.

This includes Newton's law of gravitation, and the relation between gravitational potential and field acceleration. Note that $\mathrm{d}^2\mathbf{R}/\mathrm{d}t^2$ and $\mathbf{F}/m$ are both equal to the gravitational acceleration **g** (equivalent to the inertial acceleration, so same mathematical form, but also defined as gravitational force per unit mass[8]). The negative signs are inserted since the force acts antiparallel to the displacement. The equivalent field equation in terms of mass density ϱ of the attracting mass are:

$$-\nabla \cdot \mathbf{g} = \nabla^2\Phi = 4\pi G\rho$$

which contains Gauss' law for gravity, and Poisson's equation for gravity. Newton's and Gauss' law are mathematically equivalent, and are related by the divergence theorem. Poisson's equation is obtained by taking the divergence of both sides of the previous equation. These classical equations are differential equations of motion for a test particle in the presence of a gravitational field, i.e. setting up and solving these equations allows the motion of a test mass to be determined and described.

The field around multiple particles is simply the vector sum of the fields around each individual particle. An object in such a field will experience a force that equals the vector sum of the forces it would feel in these individual fields. This is mathematically:[9]

$$\mathbf{g}_j^{(\text{net})} = \sum_{i \neq j} \mathbf{g}_i = \frac{1}{m_j} \sum_{i \neq j} \mathbf{F}_i = -G \sum_{i \neq j} m_i \frac{\hat{\mathbf{R}}_{ij}}{|\mathbf{R}_i - \mathbf{R}_j|^2} = -\sum_{i \neq j} \nabla \Phi_i$$

i.e. the gravitational field on mass m_j is the sum of all gravitational fields due to all other masses m_i, except the mass m_j itself. The unit vector $\hat{\mathbf{R}}_{ij}$ is in the direction of $\mathbf{R}i - \mathbf{R}j$.

2.2 General relativity

See also: Gravitational acceleration § General relativity and Gravitational potential § General relativity

In general relativity the gravitational field is determined by solving the Einstein field equations,[10]

$$\mathbf{G} = \frac{8\pi G}{c^4} \mathbf{T}.$$

Here $\mathbf{T}$ is the stress–energy tensor, $\mathbf{G}$ is the Einstein tensor, and c is the speed of light,

These equations are dependent on the distribution of matter and energy in a region of space, unlike Newtonian gravity, which is dependent only on the distribution of matter. The fields themselves in general relativity represent the curvature of spacetime. General relativity states that being in a region of curved space is equivalent to accelerating up the gradient of the field. By Newton's second law, this will cause an object to experience a fictitious force if it is held still with respect to the field. This is why a person will feel himself pulled down by the force of gravity while standing still on the Earth's surface. In general the gravitational fields predicted by general relativity differ in their effects only slightly from those predicted by classical mechanics, but there are a number of easily verifiable differences, one of the most well known being the bending of light in such fields.

2.3 See also

- Classical mechanics

- Gravitation

- Gravitational potential

- Newton's law of universal gravitation

- Newton's laws of motion

- Potential energy

- Speed of gravity

- Tests of general relativity

- Defining equation (physics)

2.4   Notes

[1] Richard Feynman (1970). *The Feynman Lectures on Physics Vol I*. Addison Wesley Longman. ISBN 978-0-201-02115-8.

[2] Geroch, Robert (1981). *General relativity from A to B*. University of Chicago Press. p. 181. ISBN 0-226-28864-1., Chapter 7, page 181

[3] Grøn, Øyvind; Hervik, Sigbjørn (2007). *Einstein's general theory of relativity: with modern applications in cosmology*. Springer Japan. p. 256. ISBN 0-387-69199-5., Chapter 10, page 256

[4] J. Foster, J. D. Nightingale, J. Foster, J. D. Nightingale; J. Foster, J. D. Nightingale, J. Foster, J. D. Nightingale (2006). *A short course in general relativity* (3 ed.). Springer Science & Business. p. 55. ISBN 0-387-26078-1., Chapter 2, page 55

[5] Richard Feynman (1970). *The Feynman Lectures on Physics Vol II*. Addison Wesley Longman. ISBN 978-0-201-02115-8. A "field" is any physical quantity which takes on different values at different points in space.

[6] Dynamics and Relativity, J.R. Forshaw, A.G. Smith, Wiley, 2009, ISBN 978-0-470-01460-8

[7] Encyclopaedia of Physics, R.G. Lerner, G.L. Trigg, 2nd Edition, VHC Publishers, Hans Warlimont, Springer, 2005

[8] Essential Principles of Physics, P.M. Whelan, M.J. Hodgeson, 2nd Edition, 1978, John Murray, ISBN 0-7195-3382-1

[9] Classical Mechanics (2nd Edition), T.W.B. Kibble, European Physics Series, Mc Graw Hill (UK), 1973, ISBN 0-07-084018-0.

[10] Gravitation, J.A. Wheeler, C. Misner, K.S. Thorne, W.H. Freeman & Co, 1973, ISBN 0-7167-0344-0

Chapter 3

Gravitational wave background

A possible target of gravitational wave detection experiments is a stochastic background of gravitational waves. This background is known as the **gravitational wave background** or the **stochastic background**. The detection of such a background would have a profound impact on early universe cosmology and on high-energy physics, opening up a new window and exploring very early times in the evolution of the universe, and correspondingly high energies, that will never be accessible by other means. The emission of gravitational waves from a large number of unresolved astrophysical sources can create a stochastic background of gravitational waves. For instance, sufficiently massive stars, at the final stage of their evolution, collapse to form a black hole or a neutron star. In this explosive supernova event gravitational waves are liberated. Also, in rapidly rotating neutron stars there is a whole class of instabilities driven by the emission of gravitational waves.

A stochastic gravitational wave background is also of theoretical interest. Given its stochastic nature, a coincidence of gravitational waves at a particular point could create stress-energy densities sufficient to produce an event horizon. This would produce a relativistic explanation of nonlocal effects.

Efforts to detect the gravitational wave background is ongoing. The first claimed, indirect detection of a gravitational wave background was reported on March 17, 2014 (see BICEP2),[1] which was later ruled out and confirmed to be the result of cosmic dust.[2]

3.1 See also

- Cosmic microwave background
- Cosmic neutrino background

3.2 External links

- Gravitational Wave Experiments and Early Universe Cosmology

3.3 References

[1] "BICEP2 finds first direct evidence of cosmic inflation". *Physics World*. Retrieved 18 March 2014.

[2] Cowen, Ron (2015-01-30). "Gravitational waves discovery now officially dead". *nature*. doi:10.1038/nature.2015.16830.

Chapter 4

Cosmic microwave background

"CMB" redirects here. For other uses, see CMB (disambiguation).

The **cosmic microwave background** (**CMB**) is the thermal radiation left over from the time of recombination in Big Bang cosmology. In older literature, the CMB is also variously known as cosmic microwave background radiation (CMBR) or "relic radiation." The CMB is a cosmic background radiation that is fundamental to observational cosmology because it is the oldest light in the universe, dating to the epoch of recombination. With a traditional optical telescope, the space between stars and galaxies (the *background*) is completely dark. However, a sufficiently sensitive radio telescope shows a faint background glow, almost exactly the same in all directions, that is not associated with any star, galaxy, or other object. This glow is strongest in the microwave region of the radio spectrum. The accidental discovery of CMB in 1964 by American radio astronomers Arno Penzias and Robert Wilson[1][2] was the culmination of work initiated in the 1940s, and earned the discoverers the 1978 Nobel Prize.

> *The CMB is a snapshot of the oldest light in our Universe, imprinted on the sky when the Universe was just 380,000 years old. It shows tiny temperature fluctuations that correspond to regions of slightly different densities, representing the seeds of all future structure: the stars and galaxies of today.*[3]

The CMB is well explained as radiation left over from an early stage in the development of the universe, and its discovery is considered a landmark test of the Big Bang model of the universe. When the universe was young, before the formation of stars and planets, it was denser, much hotter, and filled with a uniform glow from a white-hot fog of hydrogen plasma. As the universe expanded, both the plasma and the radiation filling it grew cooler. When the universe cooled enough, protons and electrons combined to form neutral atoms. These atoms could no longer absorb the thermal radiation, and so the universe became transparent instead of being an opaque fog.[4] Cosmologists refer to the time period when neutral atoms first formed as the *recombination epoch*, and the event shortly afterwards when photons started to travel freely through space rather than constantly being scattered by electrons and protons in plasma is referred to as photon decoupling. The photons that existed at the time of photon decoupling have been propagating ever since, though growing fainter and less energetic, since the expansion of space causes their wavelength to increase over time (and wavelength is inversely proportional to energy according to Planck's relation). This is the source of the alternative term *relic radiation*. The *surface of last scattering* refers to the set of points in space at the right distance from us so that we are now receiving photons originally emitted from those points at the time of photon decoupling.

Precise measurements of the CMB are critical to cosmology, since any proposed model of the universe must explain this radiation. The CMB has a thermal black body spectrum at a temperature of 2.72548±0.00057 K.[5] The spectral radiance dE_v/dv peaks at 160.2 GHz, in the microwave range of frequencies. (Alternatively if spectral radiance is defined as $dE_λ/dλ$ then the peak wavelength is 1.871 mm.) The glow is very nearly uniform in all directions, but the tiny residual variations show a very specific pattern, the same as that expected of a fairly uniformly distributed hot gas that has expanded to the current size of the universe. In particular, the spectral radiance at different angles of observation in the sky contains small anisotropies, or irregularities, which vary with the size of the region examined. They have been measured in detail, and match what would be expected if small thermal variations, generated by quantum fluctuations of matter in a very tiny space, had expanded to the size of the observable universe we see today. This is a very active field of study, with scientists seeking both better data (for example, the Planck spacecraft) and better interpretations of the initial conditions of expansion. Although many different processes might produce the general form of a black body spectrum, no model other than the Big Bang has yet explained the fluctuations. As a result, most cosmologists consider the Big Bang model of the universe to be the best explanation for the CMB.

The high degree of uniformity throughout the observable universe and its faint but measured anisotropy lend strong support for the Big Bang model in general and the ΛCDM ("Lambda Cold Dark Matter") model in particular. Moreover, the fluctuations are coherent on angular scales that are larger than the apparent cosmological horizon at recombination. Either such coherence is acausally fine-tuned, or cosmic inflation occurred.[6][7]

4.1 Features

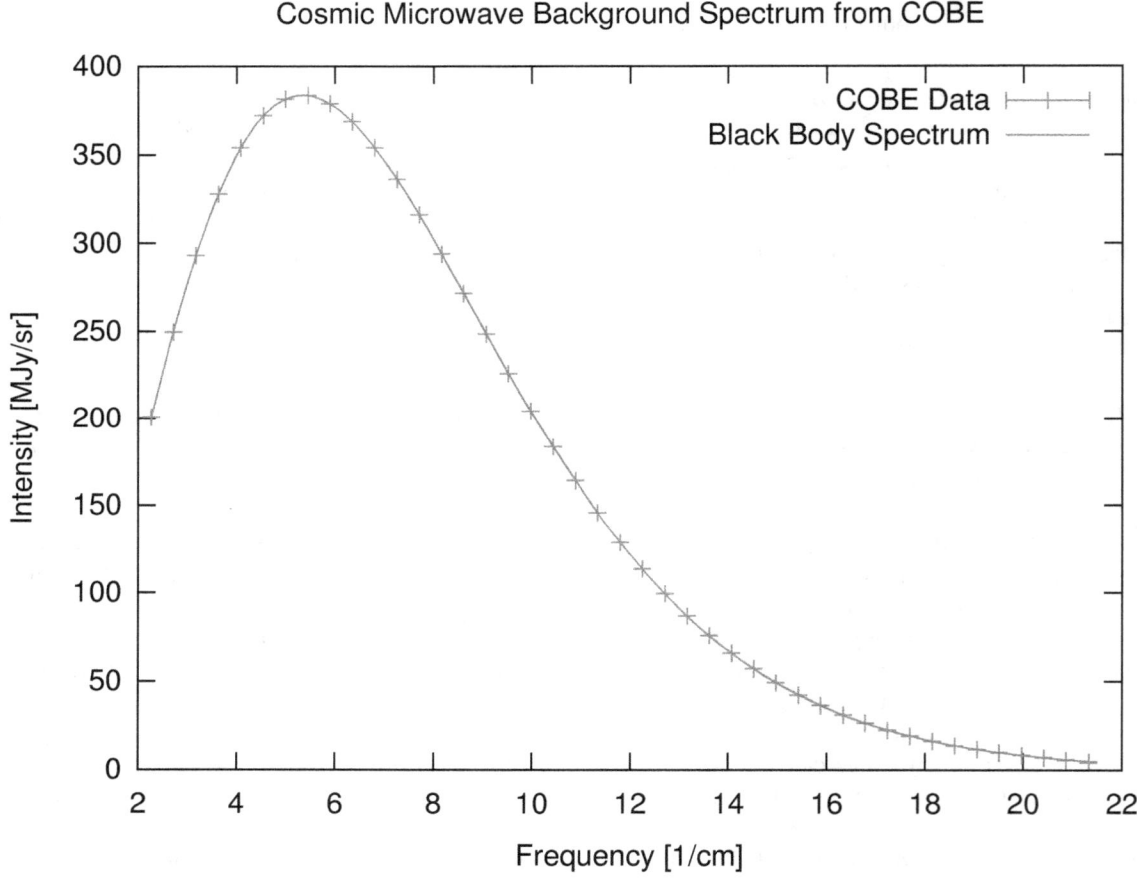

Graph of cosmic microwave background spectrum measured by the FIRAS instrument on the COBE, the most precisely measured black body spectrum in nature.[8] The error bars are too small to be seen even in an enlarged image, and it is impossible to distinguish the observed data from the theoretical curve.

The cosmic microwave background radiation is an emission of uniform, black body thermal energy coming from all parts of the sky. The radiation is isotropic to roughly one part in 100,000: the root mean square variations are only 18 μK,[9] after subtracting out a dipole anisotropy from the Doppler shift of the background radiation. The latter is caused by the peculiar velocity of the Earth relative to the comoving cosmic rest frame as the planet moves at some 371 km/s towards the constellation Leo. The CMB dipole as well as aberration at higher multipoles have been measured, consistent with galactic motion.[10]

In the Big Bang model for the formation of the universe, Inflationary Cosmology predicts that after about 10^{-37} seconds[11] the nascent universe underwent exponential growth that smoothed out nearly all inhomogeneities. The remaining inhomogeneities were caused by quantum fluctuations in the inflaton field that caused the inflation event.[12] After 10^{-6} seconds, the early universe was made up of a hot, interacting plasma of photons, electrons, and baryons. As the universe expanded, adiabatic cooling caused the energy density of the plasma to decrease until it became favorable for electrons to combine with protons, forming hydrogen atoms. This recombination event happened when the temperature was around 3000 K or when the universe was approximately 379,000 years old.[13] At this point, the photons no longer interacted with the now electrically neutral atoms and began to travel freely through space, resulting in the decoupling of matter and radiation.[14]

Thecolor temperatureof the decoupled photons has continued to diminish ever since; now down to2.7260±0.0013K, [5] it will continue to drop as the universe expands. The intensity of the radiation also corresponds to black-body radiation at 2.726 K because red-shifted black-body radiation is just like black-body radiation at a lower temperature. According to the Big Bang model, the radiation from the sky we measure today comes from a spherical surface called *the surface of last scattering*. This represents the set of locations in space at which the decoupling event is estimated to have occurred[15] and at a point in time such that the photons from that distance have just reached observers. Most of the radiation energy in the universe is in the cosmic microwave background,[16] making up a fraction of roughly 6×10^{-5} of the total density of the universe.[17]

Two of the greatest successes of the Big Bang theory are its prediction of the almost perfect black body spectrum and its detailed prediction of the anisotropies in the cosmic microwave background. The CMB spectrum has become the most precisely measured black body spectrum in nature.[8]

Density of energy for CMB is 0.25 eV/cm^{3}[18] (4.005×10^{-14} J/m^3) or (400–500 photons/cm^{3}[19]).

4.2   History

See also: Discovery of cosmic microwave background radiation

The cosmic microwave background was first predicted in 1948 by Ralph Alpher, and Robert Herman.[20][21][22] Alpher and Herman were able to estimate the temperature of the cosmic microwave background to be 5 K, though two years later they re-estimated it at 28 K. This high estimate was due to a mis-estimate of the Hubble constant by Alfred Behr, which could not be replicated and was later abandoned for the earlier estimate. Although there were several previous estimates of the temperature of space, these suffered from two flaws. First, they were measurements of the *effective* temperature of space and did not suggest that space was filled with a thermal Planck spectrum. Next, they depend on our being at a special spot at the edge of the Milky Way galaxy and they did not suggest the radiation is isotropic. The estimates would yield very different predictions if Earth happened to be located elsewhere in the universe.[23]

The 1948 results of Alpher and Herman were discussed in many physics settings through about 1955, when both left the Applied Physics Laboratory at Johns Hopkins University. The mainstream astronomical community, however, was not intrigued at the time by cosmology. Alpher and Herman's prediction was rediscovered by Yakov Zel'dovich in the early 1960s, and independently predicted by Robert Dicke at the same time. The first published recognition of the CMB radiation as a detectable phenomenon appeared in a brief paper by Soviet astrophysicists A. G. Doroshkevich and Igor Novikov, in the spring of 1964.[24] In 1964, David Todd Wilkinson and Peter Roll, Dicke's colleagues at Princeton University, began constructing a Dicke radiometer to measure the cosmic microwave background.[25] In 1964, Arno Penzias and Robert Woodrow Wilson at the Crawford Hill location of Bell Telephone Laboratories in nearby Holmdel Township, New Jersey had built a Dicke radiometer that they intended to use for radio astronomy and satellite communication experiments. On 20 May 1964 they made their first measurement clearly showing the presence of the microwave background,[26] with their instrument having an excess 4.2K antenna temperature which they could not account for. After receiving a telephone call from Crawford Hill, Dicke famously quipped: "Boys, we've been scooped."[1][27][28] A meeting between the Princeton and Crawford Hill groups determined that the antenna temperature was indeed due to the microwave background. Penzias and Wilson received the 1978 Nobel Prize in Physics for their discovery.[29]

The interpretation of the cosmic microwave background was a controversial issue in the 1960s with some proponents of the steady state theory arguing that the microwave background was the result of scattered starlight from distant galaxies.[30]

Using this model, and based on the study of narrow absorption line features in the spectra of stars, the astronomer Andrew McKellar wrote in 1941: "It can be calculated that the 'rotational temperature' of interstellar space is 2 K."[31] However, during the 1970s the consensus was established that the cosmic microwave background is a remnant of the big bang. This was largely because new measurements at a range of frequencies showed that the spectrum was a thermal, black body spectrum, a result that the steady state model was unable to reproduce.[32]

The Holmdel Horn Antenna on which Penzias and Wilson discovered the cosmic microwave background.

Harrison, Peebles, Yu and Zel'dovich realized that the early universe would have to have inhomogeneities at the level of 10^{-4} or 10^{-5}.[33][34][35] Rashid Sunyaev later calculated the observable imprint that these inhomogeneities would have on the cosmic microwave background.[36] Increasingly stringent limits on the anisotropy of the cosmic microwave background were set by ground based experiments during the 1980s. RELIKT-1, a Soviet cosmic microwave background anisotropy experiment on board the Prognoz 9 satellite (launched 1 July 1983) gave upper limits on the large-scale anisotropy. The NASA COBE mission clearly confirmed the primary anisotropy with the Differential Microwave Radiometer instrument, publishing their findings in 1992.[37][38] The team received the Nobel Prize in physics for 2006 for this discovery.

Inspired by the COBE results, a series of ground and balloon-based experiments measured cosmic microwave background anisotropies on smaller angular scales over the next decade. The primary goal of these experiments was to measure the scale of the first acoustic peak, which COBE did not have sufficient resolution to resolve. This peak corresponds to large scale density variations in the early universe that are created by gravitational instabilities, resulting in acoustical oscillations in the plasma.[39] The first peak in the anisotropy was tentatively detected by the Toco experiment and the result was confirmed by the BOOMERanG and MAXIMA experiments.[40][41][42] These measurements demonstrated that the geometry of the universe is approximately flat, rather than curved.[43] They ruled out cosmic strings as a major component of cosmic structure formation and suggested cosmic inflation was the right theory of structure formation.[44]

The second peak was tentatively detected by several experiments before being definitively detected by WMAP, which has also tentatively detected the third peak.[45] As of 2010, several experiments to improve measurements of the polarization and the microwave background on small angular scales are ongoing. These include DASI, WMAP, BOOMERanG, QUaD, Planck spacecraft, Atacama Cosmology Telescope, South Pole Telescope and the QUIET telescope.

4.2.1   Timeline

Thermal (non-microwave background) temperature predictions

- 1896 – Charles Édouard Guillaume estimates the "radiation of the stars" to be 5.6K.[46]

- 1926 – Sir Arthur Eddington estimates the non-thermal radiation of starlight in the galaxy "... by the formula $E = \sigma T^4$ the effective temperature corresponding to this density is 3.18° absolute ... black body"[47]

- 1930s – Cosmologist Erich Regener calculates that the non-thermal spectrum of cosmic rays in the galaxy has an effective temperature of 2.8 K

- 1931 – Term *microwave* first used in print: "When trials with wavelengths as low as 18 cm. were made known, there was undisguised surprise+that the problem of the micro-wave had been solved so soon." Telegraph & Telephone Journal XVII. 179/1

- 1934 – Richard Tolman shows that black-body radiation in an expanding universe cools but remains thermal

- 1938 – Nobel Prize winner (1920) Walther Nernst reestimates the cosmic ray temperature as 0.75K

- 1941 – Andrew McKellar was attempting to measure the average temperature of the interstellar medium, and used the excitation of CN doublet lines to measure that the "effective temperature of space" (the average bolometric temperature) is about 2.3 K[31][48]

- 1946 – Robert Dicke predicts "... radiation from cosmic matter" at <20 K, but did not refer to background radiation [49]

- 1946 – George Gamow calculates a temperature of 50 K (assuming a 3-billion year old universe),[50] commenting it "... is in reasonable agreement with the actual temperature of interstellar space", but does not mention background radiation.[51]

- 1953 – Erwin Finlay-Freundlich in support of his tired light theory, derives a blackbody temperature for intergalactic space of 2.3K [52] with comment from Max Born suggesting radio astronomy as the arbitrator between expanding and infinite cosmologies.

Microwave background radiation predictions

- 1946 – George Gamow calculates a temperature of 50 K (assuming a 3-billion year old universe),[50] commenting it "... is in reasonable agreement with the actual temperature of interstellar space", but does not mention background radiation.

- 1948 – Ralph Alpher and Robert Herman estimate "the temperature in the universe" at 5 K. Although they do not specifically mention microwave background radiation, it may be inferred.[53]

- 1949 – Ralph Alpher and Robert Herman re-re-estimate the temperature at 28 K.

- 1953 – George Gamow estimates 7 K.[49]

- 1956 – George Gamow estimates 6 K.[49]

- 1955 – Émile Le Roux of the Nançay Radio Observatory, in a sky survey at $\lambda = 33$ cm, reported a near-isotropic background radiation of 3 kelvins, plus or minus 2.[49]

- 1957 – Tigran Shmaonov reports that "the absolute effective temperature of the radioemission background ... is 4±3 K".[54] It is noted that the "measurements showed that radiation intensity was independent of either time or direction of observation ... it is now clear that Shmaonov did observe the cosmic microwave background at a wavelength of 3.2 cm"[55][56]

- 1960s – Robert Dicke re-estimates a microwave background radiation temperature of 40 K[49][57]

- 1964 – A. G. Doroshkevich and Igor Dmitrievich Novikov publish a brief paper suggesting microwave searches for the black-body radiation predicted by Gamow, Alpher, and Herman, where they name the CMB radiation phenomenon as detectable.[58]

- 1964–65 – Arno Penzias and Robert Woodrow Wilson measure the temperature to be approximately 3 K. Robert Dicke, James Peebles, P. G. Roll, and D. T. Wilkinson interpret this radiation as a signature of the big bang.

- 1966 – Rainer K. Sachs and Arthur M. Wolfe theoretically predict microwave background fluctuation amplitudes created by gravitational potential variations between observers and the last scattering surface (see Sachs-Wolfe effect)

- 1968 – Martin Rees and Dennis Sciama theoretically predict microwave background fluctuation amplitudes created by photons traversing time-dependent potential wells

- 1969 – R. A. Sunyaev and Yakov Zel'dovich study the inverse Compton scattering of microwave background photons by hot electrons (see Sunyaev-Zel'dovich effect)

- 1983 – Researchers from the Cambridge Radio Astronomy Group and the Owens Valley Radio Observatory first detect the Sunyaev-Zel'dovich effect from clusters of galaxies

- 1983 – RELIKT-1 Soviet CMB anisotropy experiment was launched.

- 1990 – FIRAS on the Cosmic Background Explorer (COBE) satellite measures the black body form of the CMB spectrum with exquisite precision, and shows that the microwave background has a nearly perfect black-body spectrum and thereby strongly constrains the density of the intergalactic medium.

- January 1992 – Scientists that analysed data from the RELIKT-1 report the discovery of anisotropy in the cosmic microwave background at the Moscow astrophysical seminar.[59]

- 1992 – Scientists that analysed data from COBE DMR report the discovery of anisotropy in the cosmic microwave background.[60]

- 1995 – The Cosmic Anisotropy Telescope performs the first high resolution observations of the cosmic microwave background.

- 1999 – First measurements of acoustic oscillations in the CMB anisotropy angular power spectrum from the TOCO, BOOMERANG, and Maxima Experiments. The BOOMERanG experiment makes higher quality maps at intermediate resolution, and confirms that the universe is "flat".

- 2002 – Polarization discovered by DASI.[61]

- 2003 – E-mode polarization spectrum obtained by the CBI.[62] The CBI and the Very Small Array produces yet higher quality maps at high resolution (covering small areas of the sky).

- 2003 – The WMAP spacecraft produces an even higher quality map at low and intermediate resolution of the whole sky (WMAP provides *no* high-resolution data, but improves on the intermediate resolution maps from BOOMERanG).

- 2004 – E-mode polarization spectrum obtained by the CBI.[63]

- 2004 – The Arcminute Cosmology Bolometer Array Receiver produces a higher quality map of the high resolution structure not mapped by WMAP.

- 2005 – The Arcminute Microkelvin Imager and the Sunyaev-Zel'dovich Array begin the first surveys for very high redshift clusters of galaxies using the Sunyaev-Zel'dovich effect.

- 2005 – Ralph A. Alpher is awarded the National Medal of Science for his groundbreaking work in nucleosynthesis and prediction that the universe expansion leaves behind background radiation, thus providing a model for the Big Bang theory.

- 2006 – The long-awaited three-year WMAP results are released, confirming previous analysis, correcting several points, and including polarization data.

- 2006 – Two of COBE's principal investigators, George Smoot and John Mather, received the Nobel Prize in Physics in 2006 for their work on precision measurement of the CMBR.

- 2006-2011 – Improved measurements from WMAP, new supernova surveys ESSENCE and SNLS, and baryon acoustic oscillations from SDSS and WiggleZ, continue to be consistent with the standard Lambda-CDM model.

- 2014 – On March 17, 2014, astrophysicists of the BICEP2 collaboration announced the detection of inflationary gravitational waves in the B-mode power spectrum, which if confirmed, would provide clear experimental evidence for the theory of inflation.[64][65][66][67][68][69] However, on 19 June 2014, lowered confidence in confirming the cosmic inflation findings was reported.[68][70][71]

- 2015 – On January 30, 2015, the same team of astronomers from BICEP2 withdrew the claim made on the previous year. Based on the combined data of BICEP2 and Planck, the European Space Agency announced that the signal can be entirely attributed to dust in the Milky Way.[72]

4.3   Relationship to the Big Bang

The cosmic microwave background radiation and the cosmological redshift-distance relation are together regarded as the best available evidence for the Big Bang theory. Measurements of the CMB have made the inflationary Big Bang theory the Standard Model of Cosmology.[73] The discovery of the CMB in the mid-1960s curtailed interest in alternatives such as the steady state theory.[74]

The CMB essentially confirms the Big Bang theory. In the late 1940s Alpher and Herman reasoned that if there was a big bang, the expansion of the universe would have stretched and cooled the high-energy radiation of the very early universe into the microwave region and down to a temperature of about 5 K. They were slightly off with their estimate, but they had exactly the right idea. They predicted the CMB. It took another 15 years for Penzias and Wilson to stumble into discovering that the microwave background was actually there.[75]

The CMB gives a snapshot of the universe when, according to standard cosmology, the temperature dropped enough to allow electrons and protons to form hydrogen atoms, thus making the universe transparent to radiation. When it originated some 380,000 years after the Big Bang—this time is generally known as the "time of last scattering" or the period of recombination or decoupling—the temperature of the universe was about 3000 K. This corresponds to an energy of about 0.25 eV, which is much less than the 13.6 eV ionization energy of hydrogen.[76]

Since decoupling, the temperature of the background radiation has dropped by a factor of roughly 1,100[77] due to the expansion of the universe. As the universe expands, the CMB photons are redshifted, making the radiation's temperature inversely proportional to a parameter called the universe's scale length. The temperature T_r of the CMB as a function of redshift, z, can be shown to be proportional to the temperature of the CMB as observed in the present day (2.725 K or 0.235 meV):[78]

$$T_r = 2.725(1 + z)$$

For details about the reasoning that the radiation is evidence for the Big Bang, see Cosmic background radiation of the Big Bang.

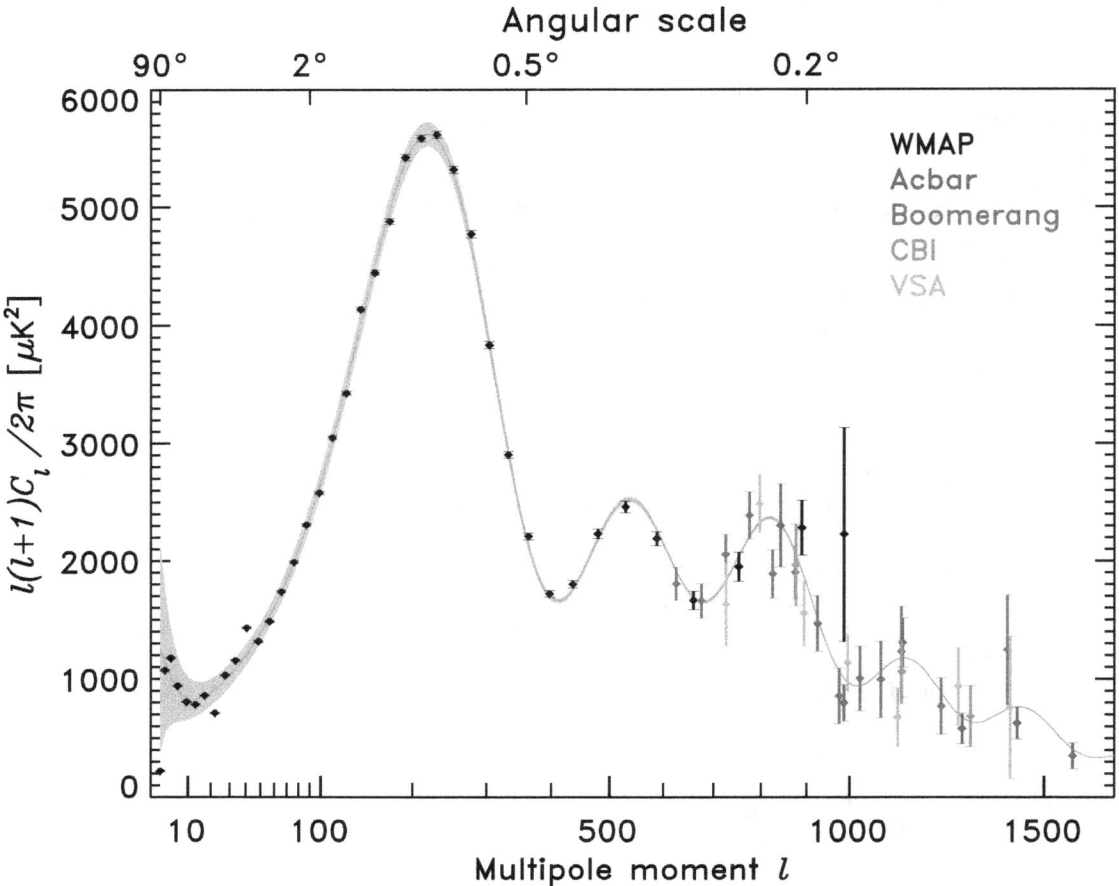

The power spectrum of the cosmic microwave background radiation temperature anisotropy in terms of the angular scale (or multipole moment). The data shown comes from the WMAP (2006), Acbar (2004) Boomerang (2005), CBI (2004), and VSA (2004) instruments. Also shown is a theoretical model (solid line).

4.3.1 Primary anisotropy

The anisotropy of the cosmic microwave background is divided into two types: primary anisotropy, due to effects which occur at the last scattering surface and before; and secondary anisotropy, due to effects such as interactions of the background radiation with hot gas or gravitational potentials, which occur between the last scattering surface and the observer.

The structure of the cosmic microwave background anisotropies is principally determined by two effects: acoustic oscillations and diffusion damping (also called collisionless damping or Silk damping). The acoustic oscillations arise because of a conflict in the photon–baryon plasma in the early universe. The pressure of the photons tends to erase anisotropies, whereas the gravitational attraction of the baryons—moving at speeds much slower than light—makes them tend to collapse to form dense haloes. These two effects compete to create acoustic oscillations which give the microwave background its characteristic peak structure. The peaks correspond, roughly, to resonances in which the photons decouple when a particular mode is at its peak amplitude.

The peaks contain interesting physical signatures. The angular scale of the first peak determines the curvature of the universe (but not the topology of the universe). The next peak—ratio of the odd peaks to the even peaks—determines the reduced baryon density.[79] The third peak can be used to get information about the dark matter density.[80]

The locations of the peaks also give important information about the nature of the primordial density perturbations. There are two fundamental types of density perturbations—called *adiabatic* and *isocurvature*. A general density perturbation is a mixture of both, and different theories that purport to explain the primordial density perturbation spectrum predict different mixtures.

- Adiabatic density perturbations

  the fractional additional density of each type of particle (baryons, photons ...) is the same. That is, if at one place there is 1% more energy in baryons than average, then at that place there is also 1% more energy in photons (and 1% more energy in neutrinos) than average. Cosmic inflation predicts that the primordial perturbations are adiabatic.

- Isocurvature density perturbations

  in each place the sum (over different types of particle) of the fractional additional densities is zero. That is, a perturbation where at some spot there is 1% more energy in baryons than average, 1% more energy in photons than average, and 2% *less* energy in neutrinos than average, would be a pure isocurvature perturbation. Cosmic strings would produce mostly isocurvature primordial perturbations.

The CMB spectrum can distinguish between these two because these two types of perturbations produce different peak locations. Isocurvature density perturbations produce a series of peaks whose angular scales (*l*-values of the peaks) are roughly in the ratio 1:3:5:..., while adiabatic density perturbations produce peaks whose locations are in the ratio 1:2:3:...[81] Observations are consistent with the primordial density perturbations being entirely adiabatic, providing key support for inflation, and ruling out many models of structure formation involving, for example, cosmic strings.

Collisionless damping is caused by two effects, when the treatment of the primordial plasma as fluid begins to break down:

- the increasing mean free path of the photons as the primordial plasma becomes increasingly rarefied in an expanding universe

- the finite depth of the last scattering surface (LSS), which causes the mean free path to increase rapidly during decoupling, even while some Compton scattering is still occurring.

These effects contribute about equally to the suppression of anisotropies at small scales, and give rise to the characteristic exponential damping tail seen in the very small angular scale anisotropies.

The depth of the LSS refers to the fact that the decoupling of the photons and baryons does not happen instantaneously, but instead requires an appreciable fraction of the age of the universe up to that era. One method of quantifying how long this process took uses the *photon visibility function* (PVF). This function is defined so that, denoting the PVF by P(t), the probability that a CMB photon last scattered between time t and t+dt is given by P(t)dt.

The maximum of the PVF (the time when it is most likely that a given CMB photon last scattered) is known quite precisely. The first-year WMAP results put the time at which P(t) is maximum as 372,000 years.[82] This is often taken as the "time" at which the CMB formed. However, to figure out how *long* it took the photons and baryons to decouple, we need a measure of the width of the PVF. The WMAP team finds that the PVF is greater than half of its maximum value (the "full width at half maximum", or FWHM) over an interval of 115,000 years. By this measure, decoupling took place over roughly 115,000 years, and when it was complete, the universe was roughly 487,000 years old.

4.3.2   Late time anisotropy

Since the CMB came into existence, it has apparently been modified by several subsequent physical processes, which are collectively referred to as late-time anisotropy, or secondary anisotropy. When the CMB photons became free to travel unimpeded, ordinary matter in the universe was mostly in the form of neutral hydrogen and helium atoms. However, observations of galaxies today seem to indicate that most of the volume of the intergalactic medium (IGM) consists of ionized material (since there are few absorption lines due to hydrogen atoms). This implies a period of reionization during which some of the material of the universe was broken into hydrogen ions.

The CMB photons are scattered by free charges such as electrons that are not bound in atoms. In an ionized universe, such charged particles have been liberated from neutral atoms by ionizing (ultraviolet) radiation. Today these free charges are at sufficiently low density in most of the volume of the universe that they do not measurably affect the CMB. However, if the IGM was ionized at very early times when the universe was still denser, then there are two main effects on the CMB:

1. Small scale anisotropies are erased. (Just as when looking at an object through fog, details of the object appear fuzzy.)

2. The physics of how photons are scattered by free electrons (Thomson scattering) induces polarization anisotropies on large angular scales. This broad angle polarization is correlated with the broad angle temperature perturbation.

Both of these effects have been observed by the WMAP spacecraft, providing evidence that the universe was ionized at very early times, at a redshift more than 17. The detailed provenance of this early ionizing radiation is still a matter of scientific debate. It may have included starlight from the very first population of stars (population III stars), supernovae when these first stars reached the end of their lives, or the ionizing radiation produced by the accretion disks of massive black holes.

The time following the emission of the cosmic microwave background—and before the observation of the first stars—is semi-humorously referred to by cosmologists as the dark age, and is a period which is under intense study by astronomers (See 21 centimeter radiation).

Two other effects which occurred between reionization and our observations of the cosmic microwave background, and which appear to cause anisotropies, are the Sunyaev–Zel'dovich effect, where a cloud of high-energy electrons scatters the radiation, transferring some of its energy to the CMB photons, and the Sachs–Wolfe effect, which causes photons from the Cosmic Microwave Background to be gravitationally redshifted or blueshifted due to changing gravitational fields.

4.4 Polarization

The cosmic microwave background is polarized at the level of a few microkelvin. There are two types of polarization, called E-modes and B-modes. This is in analogy to electrostatics, in which the electric field (*E*-field) has a vanishing curl and the magnetic field (*B*-field) has a vanishing divergence. The E-modes arise naturally from Thomson scattering in a heterogeneous plasma. The B-modes are not sourced by standard scalar type perturbations. Instead they can be created by two mechanisms: the first one is by gravitational lensing of E-modes, which has been measured by the South Pole Telescope in 2013;[83] the second one is from gravitational waves arising from cosmic inflation. Detecting the B-modes is extremely difficult, particularly as the degree of foreground contamination is unknown, and the weak gravitational lensing signal mixes the relatively strong E-mode signal with the B-mode signal.[84]

4.4.1 E-modes

E-modes were first seen in 2002 by the Degree Angular Scale Interferometer (DASI).

4.4.2 B-modes

Cosmologists predict two types of B-modes, the first generated during cosmic inflation shortly after the big bang,[85][86][87] and the second generated by gravitational lensing at later times.[88]

Primordial gravitational waves

Primordial gravitational waves are gravitational waves that could be observed in the polarisation of the cosmic microwave background and having their origin in the early universe. Models of cosmic inflation predict that such gravitational waves should appear; thus, their detection supports the theory of inflation, and their strength can confirm and exclude different models of inflation. It is the result of three things: inflationary expansion of space itself, reheating after inflation, and turbulent fluid mixing of matter and radiation. [89]

On 17 March 2014 it was announced that the BICEP2 instrument had detected the first type of B-modes, consistent with inflation and gravitational waves in the early universe at the level of $r = 0.20+0.07$
-0.05, which is the amount of power present in gravitational waves compared to the amount of power present in other scalar density perturbations in the very early universe. Had this been confirmed it would have provided strong evidence of cosmic inflation and the Big Bang,[64][65][66][67][90][91][92] but on 19 June 2014, considerably lowered confidence in confirming the findings was reported[68][68][70][70][71][71] and on 19 September 2014 new results of the Planck experiment reported that the results of BICEP2 can be fully attributed to cosmic dust.[93][94]

Paul Steinhardt is skeptical, suggesting that light scattering from cosmic dust and synchrotron radiation from electrons, both in the Milky Way Galaxy, could have caused the readings.[95]

Gravitational lensing

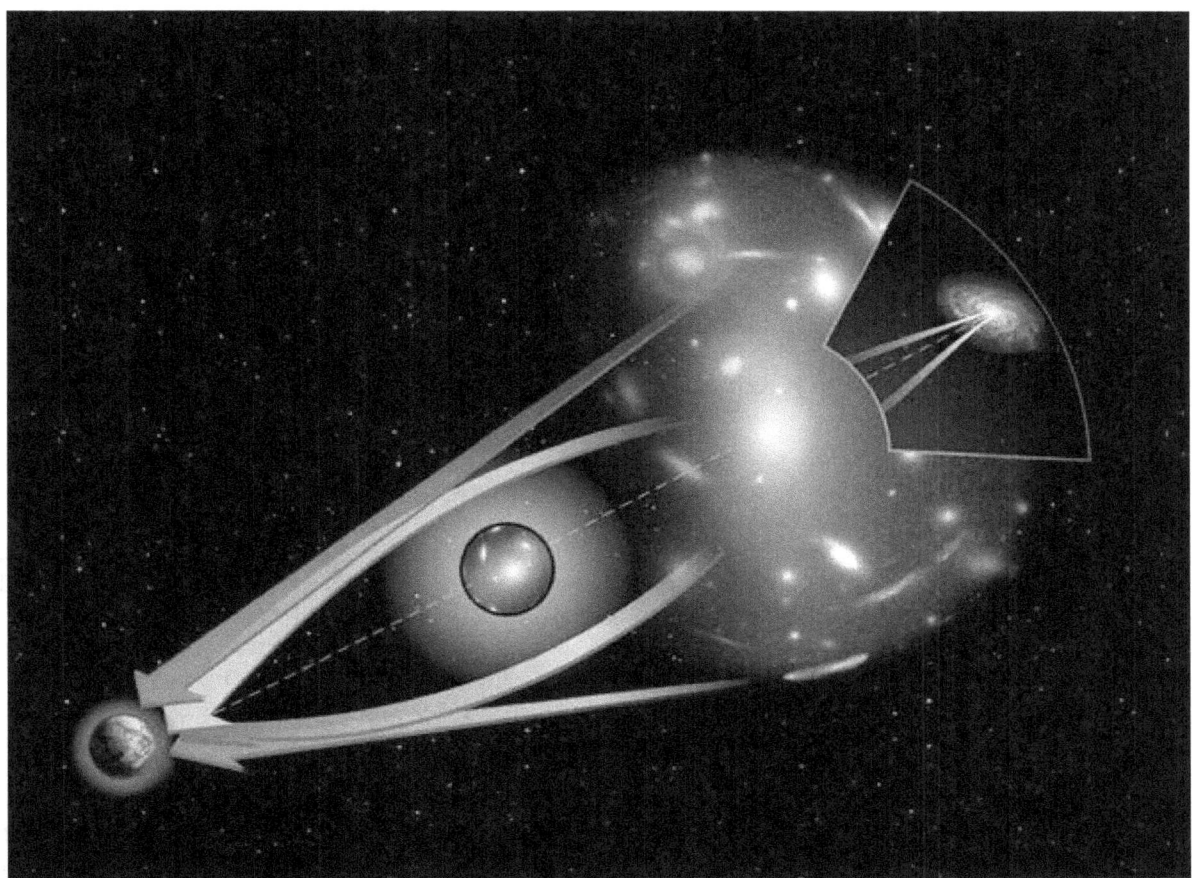

This artist's impression shows how light from the early universe is deflected by the gravitational lensing effect of massive cosmic structures forming B-modes as it travels across the universe. (Credit: ESA)

The second type of B-modes was discovered in 2013 using the South Pole Telescope with help from the Herschel Space Observatory.[96] This discovery may help test theories on the origin of the universe. Scientists are using data from the Planck mission by the European Space Agency, to gain a better understanding of these waves.[97][98][99]

In October 2014, a measurement of the B-mode polarization at 150GHz was published by the POLARBEAR experiment Compared to BICEP2, POLARBEAR focuses on a smaller patch of the sky and is less susceptible to dust effects.The team reported that POLARBEAR's measured B-mode polarization was of cosmological origin(and not just due to dust) at a 97.2% confidence level.[101]

4.5 Microwave background observations

All-sky map of the CMB, created from 9 years of WMAP data.

Main article: List of cosmic microwave background experiments

Subsequent to the discovery of the CMB, hundreds of cosmic microwave background experiments have been conducted to measure and characterize the signatures of the radiation. The most famous experiment is probably the NASA Cosmic Background Explorer (COBE) satellite that orbited in 1989–1996 and which detected and quantified the large scale anisotropies at the limit of its detection capabilities. Inspired by the initial COBE results of an extremely isotropic and homogeneous background, a series of ground- and balloon-based experiments quantified CMB anisotropies on smaller angular scales over the next decade. The primary goal of these experiments was to measure the angular scale of the first acoustic peak, for which COBE did not have sufficient resolution. These measurements were able to rule out cosmic strings as the leading theory of cosmic structure formation, and suggested cosmic inflation was the right theory. During the 1990s, the first peak was measured with increasing sensitivity and by 2000 the BOOMERanG experiment reported that the highest power fluctuations occur at scales of approximately one degree. Together with other cosmological data, these results implied that the geometry of the universe is flat. A number of ground-based interferometers provided measurements of the fluctuations with higher accuracy over the next three years, including the Very Small Array, Degree Angular Scale Interferometer (DASI), and the Cosmic Background Imager (CBI). DASI made the first detection of the polarization of the CMB and the CBI provided the first E-mode polarization spectrum with compelling evidence that it is out of phase with the T-mode spectrum.

In June 2001, NASA launched a second CMB space mission, WMAP, to make much more precise measurements of the large scale anisotropies over the full sky. WMAP used symmetric, rapid-multi-modulated scanning, rapid switching radiometers to minimize non-sky signal noise.[77] The first results from this mission, disclosed in 2003, were detailed measurements of the angular power spectrum at a scale of less than one degree, tightly constraining various cosmological parameters. The results are broadly consistent with those expected from cosmic inflation as well as various other competing theories, and are available in detail at NASA's data bank for Cosmic Microwave Background (CMB) (see links below). Although WMAP provided very accurate measurements of the large scale angular fluctuations in the CMB (structures about as broad in the sky as the moon), it did not have the angular resolution to measure the smaller scale fluctuations which had been observed by former ground-based interferometers.

A third space mission, the ESA (European Space Agency) Planck Surveyor, was launched in May 2009 and is currently performing an even more detailed investigation. Planck employs both HEMT radiometers and bolometer technology and will measure the CMB at a smaller scale than WMAP. Its detectors were trialled in the Antarctic Viper telescope as ACBAR (Arcminute Cosmology Bolometer Array Receiver) experiment—which has produced the most precise measurements at small angular scales to date—and in the Archeops balloon telescope.

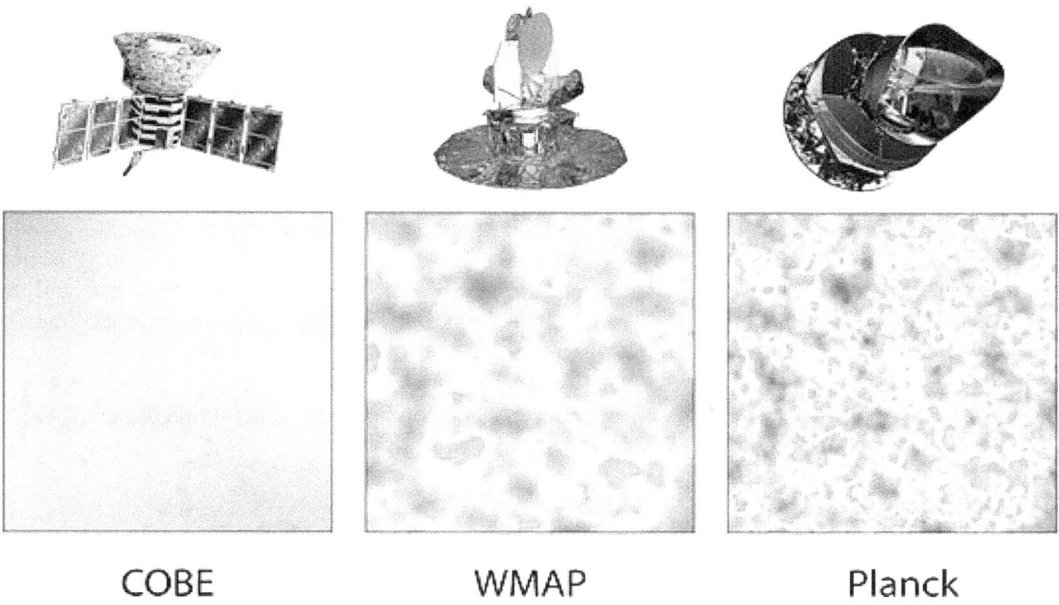

COBE   WMAP   Planck

Comparison of CMB results from COBE, WMAP and Planck – March 21, 2013.

On 21 March 2013, the European-led research team behind the Planck cosmology probe released the mission's all-sky map (565x318 jpeg, 3600x1800 jpeg) of the cosmic microwave background.[102][103] The map suggests the universe is slightly older than researchers thought. According to the map, subtle fluctuations in temperature were imprinted on the deep sky when the cosmos was about 370,000 years old. The imprint reflects ripples that arose as early, as the existence of the universe, as the first nonillionth of a second. Apparently, these ripples gave rise to the present vast cosmic web of galaxy clusters and dark matter. Based on the 2013 data, the universe contains 4.9% ordinary matter, 26.8% dark matter and 68.3% dark energy. On 5 February 2015, new data was released by the Planck mission, according to which the age of the universe is 13.799 ± 0.021 billion years old and the Hubble constant was measured to be 67.74 ± 0.46 (km/s)/Mpc.[104]

Additional ground-based instruments such as the South Pole Telescope in Antarctica and the proposed Clover Project, Atacama Cosmology Telescope and the QUIET telescope in Chile will provide additional data not available from satellite observations, possibly including the B-mode polarization.

4.6 Data reduction and analysis

Raw CMBR data from the space vehicle (i.e. WMAP) contain foreground effects that completely obscure the fine-scale structure of the cosmic microwave background. The fine-scale structure is superimposed on the raw CMBR data but is too small to be seen at the scale of the raw data. The most prominent of the foreground effects is the dipole anisotropy caused by the Sun's motion relative to the CMBR background. The dipole anisotropy and others due to Earth's annual motion relative to the Sun and numerous microwave sources in the galactic plane and elsewhere must be subtracted out to reveal the extremely tiny variations characterizing the fine-scale structure of the CMBR background.

The detailed analysis of CMBR data to produce maps, an angular power spectrum, and ultimately cosmological parameters is a complicated, computationally difficult problem. Although computing a power spectrum from a map is in principle a simple Fourier transform, decomposing the map of the sky into spherical harmonics, in practice it is hard to take the effects of noise and foreground sources into account. In particular, these foregrounds are dominated by galactic emissions such as Bremsstrahlung, synchrotron, and dust that emit in the microwave band; in practice, the galaxy has to be removed, resulting in a CMB map that is not a full-sky map. In addition, point sources like galaxies and clusters represent another source of foreground which must be removed so as not to distort the short scale structure of the CMB power spectrum.

Constraints on many cosmological parameters can be obtained from their effects on the power spectrum, and results are often calculated using Markov Chain Monte Carlo sampling techniques.

4.6.1 CMBR dipole anisotropy

From the CMB data it is seen that our local group of galaxies (the galactic cluster that includes the Solar System's Milky Way Galaxy) appears to be moving at 627±22 km/s relative to the reference frame of the CMB (also called the CMB rest frame, or the frame of reference in which there is no motion through the CMB) in the direction of galactic longitude $l = 276°±3°$, $b = 30°±3°$.[105][106] This motion results in an anisotropy of the data (CMB appearing slightly warmer in the direction of movement than in the opposite direction).[107] From a theoretical point of view, the existence of a CMB rest frame breaks Lorentz invariance even in empty space far away from any galaxy.[108] The standard interpretation of this temperature variation is a simple velocity red shift and blue shift due to motion relative to the CMB, but alternative cosmological models can explain some fraction of the observed dipole temperature distribution in the CMB.[109]

4.6.2 Low multipoles and other anomalies

With the increasingly precise data provided by WMAP, there have been a number of claims that the CMB exhibits anomalies, such as very large scale anisotropies, anomalous alignments, and non-Gaussian distributions.[110][111][112][113] The most longstanding of these is the low-l multipole controversy. Even in the COBE map, it was observed that the quadrupole ($l = 2$, spherical harmonic) has a low amplitude compared to the predictions of the Big Bang. In particular, the quadrupole and octupole ($l = 3$) modes appear to have an unexplained alignment with each other and with both the ecliptic plane and equinoxes,[114][115][116] an alignment sometimes referred to as the *axis of evil*.[111] A number of groups have suggested that this could be the signature of new physics at the greatest observable scales; other groups suspect systematic errors in the data.[117][118][119] Ultimately, due to the foregrounds and the cosmic variance problem, the greatest modes will never be as well measured as the small angular scale modes. The analyses were performed on two maps that have had the foregrounds removed as far as possible: the "internal linear combination" map of the WMAP collaboration and a similar map prepared by Max Tegmark and others.[45][77][120] Later analyses have pointed out that these are the modes most susceptible to foreground contamination from synchrotron, dust, and Bremsstrahlung emission, and from experimental uncertainty in the monopole and dipole. A full Bayesian analysis of the WMAP power spectrum demonstrates that the quadrupole prediction of Lambda-CDM cosmology is consistent with the data at the 10% level and that the observed octupole is not remarkable.[121] Carefully accounting for the procedure used to remove the foregrounds from the full sky map further reduces the significance of the alignment by ~5%.[122][123][124][125]

Recent observations with the Planck telescope, which is very much more sensitive than WMAP and has a larger angular resolution, confirm the observation of the axis of evil. Since two different instruments recorded the same anomaly, instrumental error (but not foreground contamination) appears to be ruled out.[126] Coincidence is a possible explanation, chief scientist from WMAP, Charles L. Bennett suggested coincidence and human psychology were involved, *"I do think there is a bit of a psychological effect; people want to find unusual things."* [127]

4.7 Future evolution

Assuming the universe keeps expanding and it does not suffer a Big Crunch, a Big Rip, or another similar fate, the cosmic microwave background will continue redshifting until it will no longer be detectable,[128] and will be overtaken first by the one produced by starlight, and later by the background radiation fields of processes that are assumed will take place in the far future of the universe.[129], §VD.

4.8   In popular culture

- In the *Stargate Universe* TV series, an Ancient spaceship, *Destiny*, was built to study patterns in the CMBR which indicate that the universe as we know it might have been created by some form of sentient intelligence.[130]

- In *Wheelers*, a novel by Ian Stewart & Jack Cohen, CMBR is explained as the encrypted transmissions of an ancient civilization. This allows the Jovian "blimps" to have a society older than the currently-observed age of the universe.

- In *The Three-Body Problem*, a novel by Liu Cixin, CMBR becomes observable to the naked eye due to interference from an alien civilization.

4.9   See also

- Physical cosmology

- Observational cosmology

- Gravitational wave background

- Cosmic gravitational wave background

- Observation history of galaxies

- Lambda-CDM model

- Heat death of the universe

- Computational packages for Cosmologists

4.10   References

[1] Penzias, A. A.; Wilson, R. W. (1965). "A Measurement of Excess Antenna Temperature at 4080 Mc/s". *The Astrophysical Journal* **142** (1): 419–421. Bibcode:1965ApJ...142..419P. doi:10.1086/148307.

[2] Smoot Group (28 March 1996). "The Cosmic Microwave Background Radiation". Lawrence Berkeley Lab. Retrieved 2008-12-11.

[3] "Planck reveals an almost perfect Universe". Max Planck Gesellschaft. 21 March 2013. Retrieved 2013-06-03.

[4] Kaku, M. (2014). "First Second of the Big Bang". *How the Universe Works*. Discovery Science.

[5] Fixsen, D. J. (2009). "The Temperature of the Cosmic Microwave Background". *The Astrophysical Journal* **707** (2): 916–920. arXiv:0911.1955. Bibcode:2009ApJ...707..916F. doi:10.1088/0004-637X/707/2/916.

[6] Dodelson, S. (2003). "Coherent Phase Argument for Inflation". *AIP Conference Proceedings* **689**: 184–196. arXiv:hep-ph/0309057. Bibcode:2003AIPC..689..184D. doi:10.1063/1.1627736.

[7] Baumann, D. (2011). "The Physics of Inflation" (PDF). University of Cambridge. Retrieved 2015-05-09.

[8] White, M. (1999). "Anisotropies in the CMB". *Proceedings of the Los Angeles Meeting, DPF 99*. UCLA. arXiv:astro-ph/9903232. Bibcode:1999dpf..conf.....W.

[9] Wright, E.L. (2004). "Theoretical Overview of Cosmic Microwave Background Anisotropy". In W. L. Freedman. *Measuring and Modeling the Universe*. Carnegie Observatories Astrophysics Series. Cambridge University Press. p. 291. arXiv:astro-ph/0305591. ISBN 0-521-75576-X.

[10] The Planck Collaboration, *Planck 2013 results. XXVII. Doppler boosting of the CMB: Eppur si muove*, arXiv:1303.5087, Bibcode:2014A&A...571A..27P, doi:10.1051/0004-6361/201321556

[11] Guth, A. H. (1998). *The Inflationary Universe: The Quest for a New Theory of Cosmic Origins*. Basic Books. p. 186. ISBN 978-0201328400. OCLC 35701222.

[12] Cirigliano, D.; de Vega, H.J.; Sanchez, N. G. (2005). "Clarifying inflation models: The precise inflationary potential from effective field theory and the WMAP data". *Physical Review D* **71** (10): 77–115. arXiv:astro-ph/0412634. Bibcode:2005PhRvD..71j3518C. doi:10.1103/PhysRevD.71.103518.

[13] Abbott, B. (2007). "Microwave (WMAP) All-Sky Survey". Hayden Planetarium. Retrieved 2008-01-13.

[14] Gawiser, E.; Silk, J. (2000). "The cosmic microwave background radiation". *Physics Reports*. 333–334: 245–267. arXiv:astro-ph/0002044. Bibcode:2000PhR...333..245G. doi:10.1016/S0370-1573(00)00025-9.

[15] Smoot, G. F. (2006). "Cosmic Microwave Background Radiation Anisotropies: Their Discovery and Utilization". *Nobel Lecture*. Nobel Foundation. Retrieved 2008-12-22.

[16] Hobson, M.P.; Efstathiou, G.; Lasenby, A.N. (2006). *General Relativity: An Introduction for Physicists*. Cambridge University Press. p. 388. ISBN 0-521-82951-8.

[17] Unsöld, A.; Bodo, B. (2002). *The New Cosmos, An Introduction to Astronomy and Astrophysics* (5th ed.). Springer–Verlag. p. 485. ISBN 3-540-67877-8.

[18] Confrontation of Cosmological Theories with Observational Data, M. S. Longair, page 144

[19] Cosmology II: The thermal history of the Universe, Ruth Durrer

[20] Gamow, G. (1948). "The Origin of Elements and the Separation of Galaxies". *Physical Review* **74** (4): 505–506. Bibcode:1948. doi:10.1103/PhysRev.74.505.2.

[21] Gamow, G. (1948). "The evolution of the universe". *Nature* **162** (4122): 680–682. Bibcode:1948Natur.162..680G. doi:10 PMID 18893719.

[22] Alpher, R. A.; Herman, R. C. (1948). "On the Relative Abundance of the Elements". *Physical Review* **74** (12): 1737–1742. Bibcode:1948PhRv...74.1737A. doi:10.1103/PhysRev.74.1737.

[23] Assis, A. K. T.; Neves, M. C. D. (1995). "History of the 2.7 K Temperature Prior to Penzias and Wilson" (PDF) (3). pp. 79–87. but see also Wright, E. L. (2006). "Eddington's Temperature of Space". UCLA. Retrieved 2008-12-11.

[24] Penzias, A. A. (2006). "The origin of elements" (PDF). *Nobel lecture*. Nobel Foundation. Retrieved 2006-10-04.

[25] Dicke, R. H. (1946). "The Measurement of Thermal Radiation at Microwave Frequencies". *Review of Scientific Instruments* **17** (7): 268–275. Bibcode:1946RScI...17..268D. doi:10.1063/1.1770483. PMID 20991753. This basic design for a radiometer has been used in most subsequent cosmic microwave background experiments.

[26] The Cosmic Microwave Background Radiation (Nobel Lecture) by Robert Wilson 8 Dec 1978, p. 474

[27] Dicke, R. H.; et al. (1965). "Cosmic Black-Body Radiation". *Astrophysical Journal* **142**: 414–419. Bibcode:1965ApJ...142..41 doi:10.1086/148306.

[28] The history is given in Peebles, P. J. E (1993). *Principles of Physical Cosmology*. Princeton University Press. pp. 139–148. ISBN 0-691-01933-9.

[29] "The Nobel Prize in Physics 1978". Nobel Foundation. 1978. Retrieved 2009-01-08.

[30] Narlikar, J. V.; Wickramasinghe, N. C. (1967). "Microwave Background in a Steady State Universe". *Nature* **216** (5110): 43–44. Bibcode:1967Natur.216...43N. doi:10.1038/216043a0.

[31] McKellar, A.; Kan-Mitchell, June; Conti, Peter S. (1941). "Molecular Lines from the Lowest States of Diatomic Molecules Composed of Atoms Probably Present in Interstellar Space". *Publications of the Dominion Astrophysical Observatory (Victoria, BC)* **7** (6): 251–272.

[32] Peebles, P. J. E.; et al. (1991). "The case for the relativistic hot big bang cosmology". *Nature* **352** (6338): 769–776. Bibcode:1991Natur.352..769P. doi:10.1038/352769a0.

[33] Harrison, E. R. (1970). "Fluctuations at the threshold of classical cosmology". *Physical Review D* **1** (10): 2726–2730. Bibcode:1970PhRvD...1.2726H. doi:10.1103/PhysRevD.1.2726.

[34] Peebles, P. J. E.; Yu, J. T. (1970). "Primeval Adiabatic Perturbation in an Expanding Universe". *Astrophysical Journal* **162**: 815–836. Bibcode:1970ApJ...162..815P. doi:10.1086/150713.

[35] Zeldovich, Y. B. (1972). "A hypothesis, unifying the structure and the entropy of the Universe". *Monthly Notices of the Royal Astronomical Society* **160** (7–8): 1P–4P. doi:10.1016/S0026-0576(07)80178-4.

[36] Doroshkevich, A. G.; Zel'Dovich, Y. B.; Syunyaev, R. A. (1978) [12–16 September 1977]. "Fluctuations of the microwave background radiation in the adiabatic and entropic theories of galaxy formation". In Longair, M. S.; Einasto, J. *The large scale structure of the universe; Proceedings of the Symposium*. Tallinn, Estonian SSR: Dordrecht, D. Reidel Publishing Co. pp. 393–404. Bibcode:1978IAUS...79..393S. While this is the first paper to discuss the detailed observational imprint of density inhomogeneities as anisotropies in the cosmic microwave background, some of the groundwork was laid in Peebles and Yu, above.

[37] Smooth, G. F.; et al. (1992). "Structure in the COBE differential microwave radiometer first-year maps". *Astrophysical Journal Letters* **396** (1): L1–L5. Bibcode:1992ApJ...396L...1S. doi:10.1086/186504.

[38] Bennett, C.L.; et al. (1996). "Four-Year COBE DMR Cosmic Microwave Background Observations: Maps and Basic Results". *Astrophysical Journal Letters* **464**: L1–L4. arXiv:astro-ph/9601067. Bibcode:1996ApJ...464L...1B. doi:10.1086/310075.

[39] Grupen, C.; et al. (2005). *Astroparticle Physics*. Springer. pp. 240–241. ISBN 3-540-25312-2.

[40] Miller, A. D.; et al. (1999). "A Measurement of the Angular Power Spectrum of the Microwave Background Made from the High Chilean Andes". *Astrophysical Journal* **521** (2): L79–L82. arXiv:astro-ph/9905100. Bibcode:1999ApJ...521L..79T. doi:10.1086/312197.

[41] Melchiorri, A.; et al. (2000). "A Measurement of Ω from the North American Test Flight of Boomerang". *Astrophysical Journal* **536** (2): L63–L66. arXiv:astro-ph/9911445. Bibcode:2000ApJ...536L..63M. doi:10.1086/312744.

[42] Hanany, S.; et al. (2000). "MAXIMA-1: A Measurement of the Cosmic Microwave Background Anisotropy on Angular Scales of 10'–5°".*Astrophysical Journal***545**(1): L5–L9.arXiv:astro-ph/0005123.Bibcode:2000ApJ...545L...5H.doi:10.1086/3173

[43] de Bernardis, P.; et al. (2000). "A flat Universe from high-resolution maps of the cosmic microwave background radiation". *Nature* **404** (6781): 955–959. arXiv:astro-ph/0004404. Bibcode:2000Natur.404..955D. doi:10.1038/35010035. PMID 10801117.

[44] Pogosian, L.; et al. (2003). "Observational constraints on cosmic string production during brane inflation". *Physical Review D* **68** (2): 023506. arXiv:hep-th/0304188. Bibcode:2003PhRvD..68b3506P. doi:10.1103/PhysRevD.68.023506.

[45] Hinshaw, G.; (WMAP collaboration); Bennett, C. L.; Bean, R.; Doré, O.; Greason, M. R.; Halpern, M.; Hill, R. S.; Jarosik, N.; Kogut, A.; Komatsu, E.; Limon, M.; Odegard, N.; Meyer, S. S.; Page, L.; Peiris, H. V.; Spergel, D. N.; Tucker, G. S.; Verde, L.; Weiland, J. L.; Wollack, E.; Wright, E. L.; et al. (2007). "Three-year Wilkinson Microwave Anisotropy Probe (WMAP) observations: temperature analysis". *Astrophysical Journal (Supplement Series)* **170** (2): 288–334. arXiv:astro-ph/0603451. Bibcode:2007ApJS..170..288H. doi:10.1086/513698.

[46] Guillaume, C.-É., 1896, *La Nature* 24, series 2, p. 234, cited in "History of the 2.7 K Temperature Prior to Penzias and Wilson" (PDF)

[47] Eddington, A., The Internal Constitution of the Stars, cited in "History of the 2.7 K Temperature Prior to Penzias and Wilson" (PDF)

[48] Weinberg, S. (1972). *Oxford Astronomy Encyclopedia*. John Wiley & Sons. p. 514. ISBN 0-471-92567-5.

[49] Kragh, H. (1999). *Cosmology and Controversy: The Historical Development of Two Theories of the Universe*. ISBN 0-691-00546-X. "In 1946, Robert Dicke and coworkers at MIT tested equipment that could test a cosmic microwave background of intensity corresponding to about 20K in the microwave region. However, they did not refer to such a background, but only to 'radiation from cosmic matter'. Also, this work was unrelated to cosmology and is only mentioned because it suggests that by 1950, detection of the background radiation might have been technically possible, and also because of Dicke's later role in the discovery". See also Dicke, R. H.; et al. (1946). "Atmospheric Absorption Measurements with a Microwave Radiometer". *Physical Review* **70** (5–6): 340–348. Bibcode:1946PhRv...70..340D. doi:10.1103/PhysRev.70.340.

[50] George Gamow, *The Creation Of The Universe* p.50 (Dover reprint of revised 1961 edition) ISBN 0-486-43868-6

[51] Gamow, G. (2004) [1961]. *Cosmology and Controversy: The Historical Development of Two Theories of the Universe*. Courier Dover Publications. p. 40. ISBN 978-0-486-43868-9.

[52] Erwin Finlay-Freundlich, "Ueber die Rotverschiebung der Spektrallinien" (1953) *Contributions from the Observatory, University of St. Andrews* ; no. 4, p. 96–102. Finlay-Freundlich also gave two extreme values of 1.9K and 6.0K in Finlay-Freundlich, E.: 1954, "Red shifts in the spectra of celestial bodies", Phil. Mag., Vol. 45, pp. 303–319.

[53] Helge Kragh, Cosmology and Controversy: The Historical Development of Two Theories of the Universe (1999) ISBN 0-691-00546-X. "Alpher and Herman first calculated the present temperature of the decoupled primordial radiation in 1948, when they reported a value of 5 K. Although it was not mentioned either then or in later publications that the radiation is in the microwave region, this follows immediately from the temperature ... Alpher and Herman made it clear that what they had called "the temperature in the univerese" the previous year referred to a blackbody distributed background radiation quite different from sunliight".

[54]Shmaonov, T. A. (1957). "Commentary".*Pribory i Tekhnika Experimenta*(in Russian)**1**: 83.doi:10.1016/S0890-5096(06)6073.

[55] It is noted that the "measurements showed that radiation intensity was independent of either time or direction of observation ... it is now clear that Shmaonov did observe the cosmic microwave background at a wavelength of 3.2cm"

[56] Naselsky, P. D.; Novikov, D.I.; Novikov, I. D. (2006). *The Physics of the Cosmic Microwave Background*. ISBN 0-521-85550-0.

[57] Helge Kragh, Cosmology and Controversy: The Historical Development of Two Theories of the Universe

[58] Doroshkevich, A. G.; Novikov, I.D. (1964). "Mean Density of Radiation in the Metagalaxy and Certain Problems in Relativistic Cosmology". *Soviet Physics Doklady* **9** (23): 4292–4298. Bibcode:1999EnST...33.4292W. doi:10.1021/es990537g.

[59] *Nobel Prize In Physics: Russia's Missed Opportunities*, RIA Novosti, Nov 21, 2006

[60] Sanders, R.; Kahn, J. (13 October 2006). "UC Berkeley, LBNL cosmologist George F. Smoot awarded 2006 Nobel Prize in Physics". UC Berkeley News. Retrieved 2008-12-11.

[61] Kovac, J.M.; et al. (2002). "Detection of polarization in the cosmic microwave background using DASI". *Nature* **420** (6917): 772–787. arXiv:astro-ph/0209478. Bibcode:2002Natur.420..772K. doi:10.1038/nature01269. PMID 12490941.

[62] Readhead, A. C. S.; et al. (2004). "Polarization Observations with the Cosmic Background Imager". *Science* **306** (5697): 836–844. arXiv:astro-ph/0409569. Bibcode:2004Sci...306..836R. doi:10.1126/science.1105598. PMID 15472038.

[63] A. Readhead et al., "Polarization observations with the Cosmic Background Imager", Science 306, 836-844 (2004).

[64] Staff (March 17, 2014). "BICEP2 2014 Results Release". *National Science Foundation*. Retrieved March 18, 2014.

[65] Clavin, Whitney (March 17, 2014). "NASA Technology Views Birth of the Universe". *NASA*. Retrieved March 17, 2014.

[66] Overbye, Dennis (March 17, 2014). "Space Ripples Reveal Big Bang's Smoking Gun". *The New York Times*. Retrieved March 17, 2014.

[67] Overbye, Dennis (March 24, 2014). "Ripples From the Big Bang". *New York Times*. Retrieved March 24, 2014.

[68] Ade, P.A.R. (BICEP2 Collaboration); et al. (June 19, 2014). "Detection of B-Mode Polarization at Degree Angular Scales by BICEP2" (PDF). *Physical Review Letters* **112**: 241101. arXiv:1403.3985. Bibcode:2014PhRvL.112x1101A. doi:10.1103/PhysRevLett.112.241101.PMID24996078.Retrieved June20,2014.

[69] http://www.math.columbia.edu/~{}woit/wordpress/?p=6865

[70] Overbye, Dennis (June 19, 2014). "Astronomers Hedge on Big Bang Detection Claim". *New York Times*. Retrieved June 20, 2014.

[71] Amos, Jonathan (June 19, 2014). "Cosmic inflation: Confidence lowered for Big Bang signal". *BBC News*. Retrieved June 20, 2014.

[72] Cowen, Ron (2015-01-30). "Gravitational waves discovery now officially dead". *nature*. doi:10.1038/nature.2015.16830.

[73] Scott, D. (2005). "The Standard Cosmological Model". arXiv:astro-ph/0510731 [astro-ph].

[74] Durham, Frank; Purrington, Robert D. (1983). *Frame of the universe: a history of physical cosmology*. Columbia University Press. pp. 193–209. ISBN 0-231-05393-2.

[75] Assis, A. K. T.; Paulo, São; Neves, M. C. D. (July 1995). "History of the 2.7 K Temperature Prior to Penzias and Wilson" (PDF). *Apeiron* **2** (3): 79–87.

[76] Brandenberger, Robert H. (1995). "Formation of Structure in the Universe". p. 8159.arXiv:astro-ph/9508159.Bibcode:1995

[77] Bennett, C. L.; (WMAP collaboration); Hinshaw, G.; Jarosik, N.; Kogut, A.; Limon, M.; Meyer, S. S.; Page, L.; Spergel, D. N.; Tucker, G. S.; Wollack, E.; Wright, E. L.; Barnes, C.; Greason, M. R.; Hill, R. S.; Komatsu, E.; Nolta, M. R.; Odegard, N.; Peiris, H. V.; Verde, L.; Weiland, J. L.; et al. (2003). "First-year Wilkinson Microwave Anisotropy Probe (WMAP) observations: preliminary maps and basic results". *Astrophysical Journal (Supplement Series)* **148**: 1–27. arXiv:astro-ph/0302207. Bibcode:2003ApJS..148....1B. doi:10.1086/377253. This paper warns, "the statistics of this internal linear combination map are complex and inappropriate for most CMB analyses."

[78] Noterdaeme, P.; Petitjean, P.; Srianand, R.; Ledoux, C.; López, S. (February 2011). "The evolution of the cosmic microwave background temperature. Measurements of TCMB at high redshift from carbon monoxide excitation". *Astronomy and Astrophysics* **526**: L7. arXiv:1012.3164. Bibcode:2011A&A...526L...7N. doi:10.1051/0004-6361/201016140.

[79] Wayne Hu. "Baryons and Inertia".

[80] Wayne Hu. "Radiation Driving Force".

[81] Hu, W.; White, M. (1996). "Acoustic Signatures in the Cosmic Microwave Background". *Astrophysical Journal* **471**: 30–51. arXiv:astro-ph/9602019. Bibcode:1996ApJ...471...30H. doi:10.1086/177951.

[82] WMAP Collaboration; Verde, L.; Peiris, H. V.; Komatsu, E.; Nolta, M. R.; Bennett, C. L.; Halpern, M.; Hinshaw, G.; et al. (2003). "First-Year Wilkinson Microwave Anisotropy Probe (WMAP) Observations: Determination of Cosmological Parameters". *Astrophysical Journal Supplement Series* **148** (1): 175–194. arXiv:astro-ph/0302209. Bibcode:2003ApJS..148..175S. doi:10.1086/377226.

[83] Hanson, D.; et al. (2013). "Detection of B-mode polarization in the Cosmic Microwave Background with data from the South Pole Telescope".*Physical Review Letters***111**(14).arXiv:1307.5830.Bibcode:2013PhRvL.111n1301H.doi:10.1103/PhysRev

[84] Lewis, A.; Challinor, A. (2006). "Weak gravitational lensing of the CMB". *Physics Reports* **429**: 1–65. arXiv:astro-ph/0601594. Bibcode:2006PhR...429....1L. doi:10.1016/j.physrep.2006.03.002.

[85] Seljak, U. (June 1997). "Measuring Polarization in the Cosmic Microwave Background". *Astrophysical Journal* **482**: 6–16. arXiv:astro-ph/9608131. Bibcode:1997ApJ...482....6S. doi:10.1086/304123.

[86] Seljak, U.; Zaldarriaga M. (March 17, 1997). "Signature of Gravity Waves in the Polarization of the Microwave Background". *Phys. Rev.Lett.***78**(11): 2054–2057.arXiv:astro-ph/9609169.Bibcode:1997PhRvL..78.2054S.doi:10.1103/PhysRevLett.78.

[87] Kamionkowski, M.; Kosowsky A. & Stebbins A. (March 17, 1997). "A Probe of Primordial Gravity Waves and Vorticity". *Phys. Rev.Lett.* **78** (11): 2058–2061. arXiv:astro-ph/9609132. Bibcode:1997PhRvL..78.2058K. doi:10.1103/PhysRevLett.78.2058.

[88] Zaldarriaga, M.; Seljak U. (July 15, 1998). "Gravitational lensing effect on cosmic microwave background polarization". *Physical Review D.* 2 **58**. arXiv:astro-ph/9803150. Bibcode:1998PhRvD..58b3003Z. doi:10.1103/PhysRevD.58.023003.

[89] "Scientists Report Evidence for Gravitational Waves in Early Universe". Retrieved 2007-06-20.

[90] "Gravitational waves: have US scientists heard echoes of the big bang?". The Guardian. 2014-03-14. Retrieved 2014-03-14.

[91] 'BICEP2 I: Detection Of B-mode Polarization at Degree Angular Scales' on arXiv

[92] "Space Ripples Reveal Big Bang's Smoking Gun". March 17, 2014.

[93] Planck Collaboration Team (19 September 2014). "Planck intermediate results. XXX. The angular power spectrum of polarized dust emission at intermediate and high Galactic latitudes". arXiv:1409.5738.

[94] Overbye, Dennis (22 September 2014). "Study Confirms Criticism of Big Bang Finding". *New York Times*. Retrieved 22 September 2014.

[95] "Big Bang research blunder leaves multiverse theory in ruins, theoretical physicist claims". *The Independent*.

[96] "Polarization detected in Big Bang's echo". *Nature News & Comment*.

[97] ESA Planck (Oct 22, 2013). "Planck Space Mission". Retrieved Oct 23, 2013.

[98] NASA/Jet Propulsion Laboratory (October 22, 2013). "Long-sought pattern of ancient light detected". *ScienceDaily*. Retrieved October 23, 2013.

[99] Hanson, D.; et al. (Sep 30, 2013). "Detection of B-Mode Polarization in the Cosmic Microwave Background with Data from the South Pole Telescope". *Physical Review Letters*. 14 **111**. arXiv:1307.5830. Bibcode:2013PhRvL.111n1301H. doi:10.1103/PhysRevLett.111.141301.

[100] The Polarbear Collaboration (October 2014). "A Measurement of the Cosmic Microwave Background B-Mode Polarization Power Spectrum at Sub-Degree Scales with POLARBEAR" (PDF). *The Astrophysical Journal* **794**: 171. arXiv:1403.2369. Bibcode:2014ApJ...794..171T. doi:10.1088/0004-637X/794/2/171. Retrieved November 16, 2014.

[101] "POLARBEAR project offers clues about origin of universe's cosmic growth spurt". *Christian Science Monitor*. October 21, 2014.

[102] Clavin, Whitney; Harrington, J.D. (21 March 2013). "Planck Mission Brings Universe Into Sharp Focus". *NASA*. Retrieved 21 March 2013.

[103] Staff (21 March 2013). "Mapping the Early Universe". *New York Times*. Retrieved 23 March 2013.

[104] Planck Collaboration (2015). "Planck 2015 results. XIII. Cosmological parameters (See Table 4 on page 31 of pfd).". arXiv:1502.01589. Bibcode:2015arXiv150201589P.

[105] Kogut, A.; Lineweaver, C.; Smoot, G. F.; Bennett, C. L.; Banday, A.; Boggess, N. W.; Cheng, E. S.; De Amici, G.; Fixsen, D. J.; Hinshaw, G.; Jackson, P. D.; Janssen, M.; Keegstra, P.; Loewenstein, K.; Lubin, P.; Mather, J. C.; Tenorio, L.; Weiss, R.; Wilkinson, D. T.; Wright, E. L. (1993). "Dipole Anisotropy in the COBE Differential Microwave Radiometers First-Year Sky Maps". *Astrophysical Journal* **419**: 1–6. arXiv:astro-ph/9312056. Bibcode:1993ApJ...419....1K. doi:10.1086/173453.

[106] Aghanim, N.; Armitage-Caplan, C.; et al. (2013). "Planck 2013 results. XXVII. Doppler boosting of the CMB: Eppur si muove". *Astronomy & Astrophysics* **571** (27): A27. arXiv:1303.5087. Bibcode:2014A&A...571A..27P. doi:10.1051/0004-6361/201321556.

[107] http://antwrp.gsfc.nasa.gov/apod/ap090906.html

[108] http://iopscience.iop.org/1126-6708/2005/07/029/

[109] Inoue, K. T.; Silk, J. (2007). "Local Voids as the Origin of Large-Angle Cosmic Microwave Background Anomalies: The Effect of a Cosmological Constant". *Astrophysical Journal* **664** (2): 650–659. arXiv:astro-ph/0612347. Bibcode:2007ApJ...664..650I. doi:10.1086/517603.

[110] Rossmanith, G.; Räth, C.; Banday, A. J.; Morfill, G. (2009). "Non-Gaussian Signatures in the five-year WMAP data as identified with isotropic scaling indices". *Monthly Notices of the Royal Astronomical Society* **399** (4): 1921–1933. arXiv:0905.2854. Bibcode:2009MNRAS.399.1921R. doi:10.1111/j.1365-2966.2009.15421.x.

[111] Schild, R. E.; Gibson, C. H. (2008). "Goodness in the Axis of Evil". arXiv:0802.3229 [astro-ph].

[112] Bernui, A.; Mota, B.; Rebouças, M. J.; Tavakol, R. (2005). "Mapping the large-scale anisotropy in the WMAP data". *Astronomy and Astrophysics* **464** (2): 479–485. arXiv:astro-ph/0511666. Bibcode:2007A&A...464..479B. doi:10.1051/0004-6361:20065585.

[113] Jaffe, T.R.; Banday, A. J.; Eriksen, H. K.; Górski, K. M.; Hansen, F. K. (2005). "Evidence of vorticity and shear at large angular scales in the WMAP data: a violation of cosmological isotropy?". *The Astrophysical Journal* **629**: L1–L4. arXiv:astro-ph/0503213. Bibcode:2005ApJ...629L...1J. doi:10.1086/444454.

[114] de Oliveira-Costa, A.; Tegmark, Max; Zaldarriaga, Matias; Hamilton, Andrew (2004). "The significance of the largest scale CMB fluctuations in WMAP". *Physical Review D* **69** (6): 063516. arXiv:astro-ph/0307282. Bibcode:2004PhRvD..69f3516D. doi:10.1103/PhysRevD.69.063516.

[115] Schwarz, D. J.; Starkman, Glenn D.; et al. (2004). "Is the low-l microwave background cosmic?". *Physical Review Letters* **93** (22): 221301. arXiv:astro-ph/0403353. Bibcode:2004PhRvL..93v1301S. doi:10.1103/PhysRevLett.93.221301.

[116] Bielewicz, P.; Gorski, K. M.; Banday, A. J. (2004). "Low-order multipole maps of CMB anisotropy derived from WMAP". *Monthly Notices of the Royal Astronomical Society* **355** (4): 1283–1302. arXiv:astro-ph/0405007. Bibcode:2004MNRAS.355.1 doi:10.1111/j.1365-2966.2004.08405.x.

[117] Liu, Hao; Li, Ti-Pei (2009). "Improved CMB Map from WMAP Data". arXiv:0907.2731v3 [astro-ph].

[118]Sawangwit, Utane; Shanks, Tom (2010). "Lambda-CDM and the WMAP Power Spectrum Beam Profile Sensitivity".arXiv:1006 [astro-ph].

[119] Liu, Hao; et al. (2010). "Diagnosing Timing Error in WMAP Data". arXiv:1009.2701v1 [astro-ph].

[120] Tegmark, M.; de Oliveira-Costa, A.; Hamilton, A. (2003). "A high resolution foreground cleaned CMB map from WMAP". *Physical Review D* **68** (12): 123523. arXiv:astro-ph/0302496. Bibcode:2003PhRvD..68l3523T. doi:10.1103/PhysRevD.68.123 523.This paper states, "Not surprisingly,the two most contaminated multipoles are[the quadrupole and octupole],which most closelytrace the galactic plane morphology."

[121] O'Dwyer, I.; Eriksen, H. K.; Wandelt, B. D.; Jewell, J. B.; Larson, D. L.; Górski, K. M.; Banday, A. J.; Levin, S.; Lilje, P. B. (2004). "Bayesian Power Spectrum Analysis of the First-Year Wilkinson Microwave Anisotropy Probe Data". *Astrophysical Journal Letters* **617** (2): L99–L102. arXiv:astro-ph/0407027. Bibcode:2004ApJ...617L..99O. doi:10.1086/427386.

[122] Slosar, A.; Seljak, U. (2004). "Assessing the effects of foregrounds and sky removal in WMAP". *Physical Review D* **70** (8): 083002. arXiv:astro-ph/0404567. Bibcode:2004PhRvD..70h3002S. doi:10.1103/PhysRevD.70.083002.

[123] Bielewicz, P.; Eriksen, H. K.; Banday, A. J.; Górski, K. M.; Lilje, P. B. (2005). "Multipole vector anomalies in the first-year WMAP data: a cut-sky analysis". *Astrophysical Journal* **635** (2): 750–60. arXiv:astro-ph/0507186. Bibcode:2005ApJ...635.. 750B.doi:10.1086/497263.

[124] Copi, C.J.; Huterer, Dragan; Schwarz, D. J.; Starkman, G. D. (2006). "On the large-angle anomalies of the microwave sky". *Monthly Notices of the Royal Astronomical Society* **367**: 79–102. arXiv:astro-ph/0508047. Bibcode:2006MNRAS.367...79C. doi:10.1111/j.1365-2966.2005.09980.x.

[125] de Oliveira-Costa, A.; Tegmark, M. (2006). "CMB multipole measurements in the presence of foregrounds". *Physical Review D* **74** (2): 023005. arXiv:astro-ph/0603369. Bibcode:2006PhRvD..74b3005D. doi:10.1103/PhysRevD.74.023005.

[126] Planck shows almost perfect cosmos – plus axis of evil

[127] Found: Hawking's initials written into the universe

[128] Krauss, Lawrence M.; Scherrer, Robert J. (2007). "The return of a static universe and the end of cosmology". *General Relativity and Gravitation* **39** (10): 1545–1550. arXiv:0704.0221. Bibcode:2007GReGr..39.1545K. doi:10.1007/s10714-007-0472-9.

[129] Adams, Fred C.; Laughlin, Gregory (1997). "A dying universe: The long-term fate and evolution of astrophysical objects". *Reviews of Modern Physics***69**(2): 337–372.arXiv:astro-ph/9701131.Bibcode:1997RvMP...69..337A.doi:10.1103/RevMod

[130] Cosmic Rebirth Encoded in Background Radiation?

4.11   External links

- Student Friendly Intro to the CMB A pedagogic, step-by-step introduction to the cosmic microwave background power spectrum analysis suitable for those with an undergraduate physics background. More in depth than typical online sites. Less dense than cosmology texts.

- CMBR Theme on arxiv.org

- Audio: Fraser Cain and Dr. Pamela Gay – Astronomy Cast. The Big Bang and Cosmic Microwave Background – October 2006

- Visualization of the CMB data from the Planck mission

- Copeland, Ed. "CMBR: Cosmic Microwave Background Radiation". *Sixty Symbols*. Brady Haran for the University of Nottingham.

Chapter 5

Graviton

This article is about the hypothetical particle. For other uses, see Graviton (disambiguation).

In physics, the **graviton** is a hypothetical elementary particle that mediates the force of gravitation in the framework of quantum field theory. If it exists, the graviton is expected to be massless (because the gravitational force appears to have unlimited range) and must be a spin-2 boson. The spin follows from the fact that the source of gravitation is the stress–energy tensor, a second-rank tensor (compared to electromagnetism's spin-1 photon, the source of which is the four-current, a first-rank tensor). Additionally, it can be shown that any massless spin-2 field would give rise to a force indistinguishable from gravitation, because a massless spin-2 field must couple to (interact with) the stress–energy tensor in the same way that the gravitational field does. Seeing as the graviton is hypothetical, its discovery would unite quantum theory with gravity.[4] This result suggests that, if a massless spin-2 particle is discovered, it must be the graviton, so that the only experimental verification needed for the graviton may simply be the discovery of a massless spin-2 particle.[5]

5.1 Theory

The four other known forces of nature are mediated by elementary particles: electromagnetism by the photon, the strong interaction by the gluons, the Higgs field by the Higgs Boson, and the weak interaction by the W and Z bosons. The hypothesis is that the gravitational interaction is likewise mediated by an – as yet undiscovered – elementary particle, dubbed as *the graviton*. In the classical limit, the theory would reduce to general relativity and conform to Newton's law of gravitation in the weak-field limit.[6][7][8]

5.1.1 Gravitons and renormalization

When describing graviton interactions, the classical theory (i.e., the tree diagrams) and semiclassical corrections (one-loop diagrams) behave normally, but Feynman diagrams with two (or more) loops lead to ultraviolet divergences; that is, infinite results that cannot removed because the quantized general relativity is not renormalizable, unlike quantum electrodynamics. That is, the usual ways physicists calculate the probability that a particle will emit or absorb a graviton give nonsensical answers and the theory loses its predictive power. These problems, together with some conceptual puzzles, led many physicists to believe that a theory more complete than quantized general relativity must describe the behavior near the Planck scale.

5.1.2 Comparison with other forces

Unlike the force carriers of the other forces, gravitation plays a special role in general relativity in defining the spacetime in which events take place. In some descriptions, matter modifies the 'shape' of spacetime itself, and gravity is a result of this shape, an idea which at first glance may appear hard to match with the idea of a force acting between particles.[9]

Because the diffeomorphism invariance of the theory does not allow any particular space-time background to be singled out as the "true" space-time background, general relativity is said to be background independent. In contrast, the Standard Model is *not* background independent, with Minkowski space enjoying a special status as the fixed background space-time.[10] A theory of quantum gravity is needed in order to reconcile these differences.[11] Whether this theory should be background independent is an open question. The answer to this question will determine our understanding of what specific role gravitation plays in the fate of the universe.[12]

5.1.3   Gravitons in speculative theories

String theory predicts the existence of gravitons and their well-defined interactions. A graviton in perturbative string theory is a closed string in a very particular low-energy vibrational state. The scattering of gravitons in string theory can also be computed from the correlation functions in conformal field theory, as dictated by the AdS/CFT correspondence, or from matrix theory.

A feature of gravitons in string theory is that, as closed strings without endpoints, they would not be bound to branes and could move freely between them. If we live on a brane (as hypothesized by brane theories) this "leakage" of gravitons from the brane into higher-dimensional space could explain why gravitation is such a weak force, and gravitons from other branes adjacent to our own could provide a potential explanation for dark matter. However, if gravitons were to move completely freely between branes this would dilute gravity too much, causing a violation of Newton's inverse square law. To combat this, Lisa Randall found that a three-brane (such as ours) would have a gravitational pull of its own, preventing gravitons from drifting freely, possibly resulting in the diluted gravity we observe while roughly maintaining Newton's inverse square law.[13] See brane cosmology.

A theory by Ahmed Farag Ali and Saurya Das adds quantum mechanical corrections (using Bohm trajectories) to general relativistic geodesics. If gravitons are given a small but non-zero mass, it could explain the cosmological constant without need for dark energy and solve the smallness problem.[14]

5.2   Experimental observation

Unambiguous detection of individual gravitons, though not prohibited by any fundamental law, is impossible with any physically reasonable detector.[15] The reason is the extremely low cross section for the interaction of gravitons with matter. For example, a detector with the mass of Jupiter and 100% efficiency, placed in close orbit around a neutron star, would only be expected to observe one graviton every 10 years, even under the most favorable conditions. It would be impossible to discriminate these events from the background of neutrinos, since the dimensions of the required neutrino shield would ensure collapse into a black hole.[15]

However, experiments to detect gravitational waves, which may be viewed as coherent states of many gravitons, are underway (such as LIGO and VIRGO). Although these experiments cannot detect individual gravitons, they might provide information about certain properties of the graviton.[16] For example, if gravitational waves were observed to propagate slower than c (the speed of light in a vacuum), that would imply that the graviton has mass (however, gravitational waves must propagate slower than "c" in a region with non-zero mass density if they are to be detectable).[17] Astronomical observations of the kinematics of galaxies, especially the galaxy rotation problem and modified Newtonian dynamics, might point toward gravitons having non-zero mass.[18]

5.3   Difficulties and outstanding issues

Most theories containing gravitons suffer from severe problems. Attempts to extend the Standard Model or other quantum field theories by adding gravitons run into serious theoretical difficulties at high energies (processes involving energies close to or above the Planck scale) because of infinities arising due to quantum effects (in technical terms, gravitation is nonrenormalizable). Since classical general relativity and quantum mechanics seem to be incompatible at such energies, from a theoretical point of view, this situation is not tenable. One possible solution is to replace particles with strings.

String theories are quantum theories of gravity in the sense that they reduce to classical general relativity plus field theory at low energies, but are fully quantum mechanical, contain a graviton, and are believed to be mathematically consistent.[19]

5.4 See also

- Gravitomagnetism

- Gravitational wave

- Planck mass

- Gravitation

- Static forces and virtual-particle exchange

- Multiverse

- Gravitino

5.5 References

[1] G is used to avoid confusion with gluons (symbol g)

[2] Rovelli, C. (2001). "Notes for a brief history of quantum gravity". arXiv:gr-qc/0006061 [gr-qc].

[3] Blokhintsev, D. I.; Gal'perin, F. M. (1934). "Gipoteza neitrino i zakon sokhraneniya energii" [Neutrino hypothesis and conservation of energy]. *Pod Znamenem Marxisma* (in Russian) **6**: 147–157.

[4] Lightman, A. P.; Press, W. H.; Price, R. H.; Teukolsky, S. A. (1975). "Problem 12.16". *Problem book in Relativity and Gravitation*. Princeton University Press. ISBN 0-691-08162-X.

[5] For a comparison of the geometric derivation and the (non-geometric) spin-2 field derivation of general relativity, refer to box 18.1 (and also 17.2.5) of Misner, C. W.; Thorne, K. S.; Wheeler, J. A. (1973). *Gravitation*. W. H. Freeman. ISBN 0-7167-0344-0.

[6] Feynman, R. P.; Morinigo, F. B.; Wagner, W. G.; Hatfield, B. (1995). *Feynman Lectures on Gravitation*. Addison-Wesley. ISBN 0-201-62734-5.

[7] Zee, A. (2003). *Quantum Field Theory in a Nutshell*. Princeton University Press. ISBN 0-691-01019-6.

[8] Randall, L. (2005). *Warped Passages: Unraveling the Universe's Hidden Dimensions*. Ecco Press. ISBN 0-06-053108-8.

[9] See the other articles on General relativity, Gravitational field, Gravitational wave, etc

[10] Colosi, D.; et al. (2005). "Background independence in a nutshell: The dynamics of a tetrahedron". *Classical and Quantum Gravity* **22** (14): 2971. arXiv:gr-qc/0408079. Bibcode:2005CQGra..22.2971C. doi:10.1088/0264-9381/22/14/008.

[11] Witten, E. (1993). "Quantum Background Independence In String Theory". arXiv:hep-th/9306122 [hep-th].

[12] Smolin, L. (2005). "The case for background independence". arXiv:hep-th/0507235 [hep-th].

[13] Kaku, Michio (2006). *Parallel Worlds - The science of alternative universes and our future in the Cosmos*. pp. 218–221.

[14] Ali, Ahmed Farang (2014). "Cosmology from quantum potential". *Physical Letters B* **741**: 276–279. arXiv:1404.3093v3. doi:10.1016/j.physletb.2014.12.057.

[15] Rothman, T.; Boughn, S. (2006). "Can Gravitons be Detected?". *Foundations of Physics* **36** (12): 1801–1825. arXiv:gr-qc/0601043. Bibcode:2006FoPh...36.1801R. doi:10.1007/s10701-006-9081-9.

[16] Dyson, Freeman (8 October 2013). "Is a graviton detectable?". *International Journal of Modern Physics A* **28** (25): 1330041-1-1330035–14. Bibcode:2013IJMPA..2830041D. doi:10.1142/S0217751X1330041X.

[17] Will, C. M. (1998). "Bounding the mass of the graviton using gravitational-wave observations of inspiralling compact binaries". *Physical Review D* **57** (4): 2061–2068. arXiv:gr-qc/9709011. Bibcode:1998PhRvD..57.2061W. doi:10.1103/PhysRevD.57.2061.

[18] Trippe, S. (2013), "A Simplified Treatment of Gravitational Interaction on Galactic Scales", J. Kor. Astron. Soc. **46**, 41. arXiv:1211.4692

[19] Sokal, A. (July 22, 1996). "Don't Pull the String Yet on Superstring Theory". *The New York Times*. Retrieved March 26, 2010.

5.6   External links

-

- Graviton on *In Our Time* at the BBC. (listen now)

Chapter 6

Gravitoelectromagnetism

This article is about the gravitational analog of electromagnetism as a whole. For the specific gravitational analog of magnetism, see frame-dragging.

Gravitoelectromagnetism, abbreviated **GEM**, refers to a set of formal analogies between the equations for electromagnetism

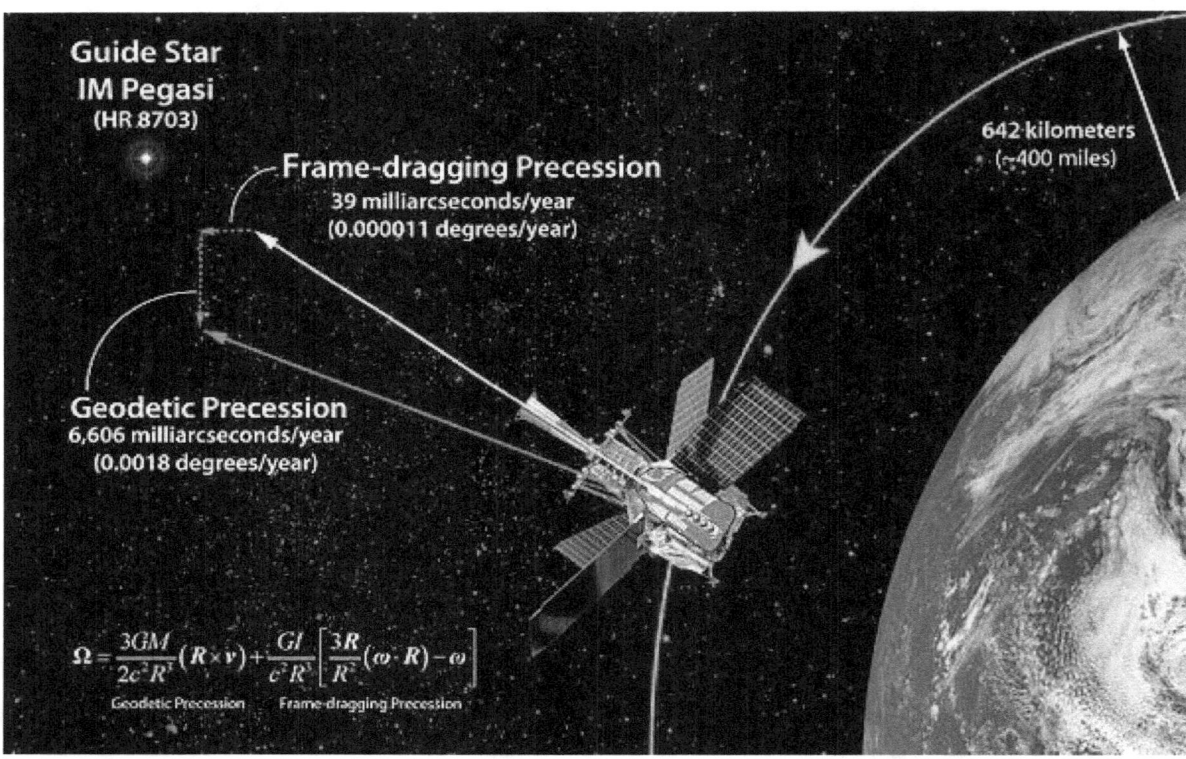

Diagram regarding the confirmation of gravitomagnetism by Gravity Probe B

and relativistic gravitation; specifically: between Maxwell's field equations and an approximation, valid under certain conditions, to the Einstein field equations for general relativity. **Gravitomagnetism** is a widely used term referring specifically to the kinetic effects of gravity, in analogy to the magnetic effects of moving electric charge. The most common version of GEM is valid only far from isolated sources, and for slowly moving test particles.

The analogy and equations differing only by some small factors were first published in 1893, before general relativity, by Oliver Heaviside as a separate theory expanding Newton's law.[1]

6.1   Background

This approximate reformulation of gravitation as described by general relativity in the weak field limit makes an apparent field appear in a frame of reference different from that of a freely moving inertial body. This apparent field may be described by two components that act respectively like the electric and magnetic fields of electromagnetism, and by analogy these are called the *gravitoelectric* and *gravitomagnetic* fields, since these arise in the same way around a mass that a moving electric charge is the source of electric and magnetic fields. The main consequence of the *gravitomagnetic* field, or velocity-dependent acceleration, is that a moving object near a rotating massive object will experience acceleration not predicted by a purely Newtonian (gravitoelectric) gravity field. More subtle predictions, such as induced rotation of a falling object and precession of a spinning object are among the last basic predictions of general relativity to be directly tested.

Indirect validations of gravitomagnetic effects have been derived from analyses of relativistic jets. Roger Penrose had proposed a frame dragging mechanism for extracting energy and momentum from rotating black holes.[2] Reva Kay Williams, University of Florida, developed a rigorous proof that validated Penrose's mechanism.[3] Her model showed how the Lense–Thirring effect could account for the observed high energies and luminosities of quasars and active galactic nuclei; the collimated jets about their polar axis; and the asymmetrical jets (relative to the orbital plane).[4] All of those observed properties could be explained in terms of gravitomagnetic effects.[5] Williams' application of Penrose's mechanism can be applied to black holes of any size.[6] Relativistic jets can serve as the largest and brightest form of validations for gravitomagnetism.

A group at Stanford University is currently analyzing data from the first direct test of GEM, the Gravity Probe B satellite experiment, to see if they are consistent with gravitomagnetism. The Apache Point Observatory Lunar Laser-ranging Operation also plans to observe gravitomagnetism effects.

6.2   Equations

According to general relativity, the gravitational field produced by a rotating object (or any rotating mass–energy) can, in a particular limiting case, be described by equations that have the same form as in classical electromagnetism. Starting from the basic equation of general relativity, the Einstein field equation, and assuming a weak gravitational field or reasonably flat spacetime, the gravitational analogs to Maxwell's equations for electromagnetism, called the "GEM equations", can be derived. GEM equations compared to Maxwell's equations in SI units are:[8][9]

where:

- $\mathbf{E}_g$ is the static gravitational field (conventional gravity, also called *gravitoelectric* in analogous usage) in $\text{m}\cdot\text{s}^{-2}$;

- $\mathbf{E}$ is the electric field;

- $\mathbf{B}_g$ is the gravitomagnetic field in s^{-1};

- $\mathbf{B}$ is the magnetic field;

- ϱ_g is mass density in $\text{kg}\cdot\text{m}^{-3}$;

- ϱ is charge density:

- $\mathbf{J}_g$ is mass current density or mass flux ($\mathbf{J}_g = \varrho_g \mathbf{v}_\rho$, where $\mathbf{v}_\rho$ is the velocity of the mass flow generating the gravitomagnetic field) in $\text{kg}\cdot\text{m}^{-2}\cdot\text{s}^{-1}$;

- $\mathbf{J}$ is electric current density;

- G is the gravitational constant in $\text{m}^3\cdot\text{kg}^{-1}\cdot\text{s}^{-2}$;

- ε_0 is the vacuum permittivity;

- c is the speed of propagation of gravity (which is equal to the speed of light according to general relativity) in $\text{m}\cdot\text{s}^{-1}$.

6.2.1 Lorentz force

For a test particle whose mass m is "small", in a stationary system, the net (Lorentz) force acting on it due to a GEM field is described by the following GEM analog to the Lorentz force equation:

where:

- v is the velocity of the test particle;

- m is the mass of the test particle;

- q is the electric charge of the test particle.

6.2.2 Poynting vector

The GEM Poynting vector compared to the electromagnetic Poynting vector is given by[10]

6.2.3 Scaling of fields

The literature does not adopt a consistent scaling for the gravitoelectric and gravitomagnetic fields, making comparison tricky. For example, to obtain agreement with Mashhoon's writings, all instances of B_g in the GEM equations must be multiplied by $-1/2c$ and E_g by -1. These factors variously modify the analogues of the equations for the Lorentz force. No scaling choice allows all the GEM and EM equations to be perfectly analogous. The discrepancy in the factors arises because the source of the gravitational field is the second order stress–energy tensor, as opposed to the source of the electromagnetic field being the first order four-current tensor. This difference becomes clearer when one compares non-invariance of relativistic mass to electric charge invariance. This can be traced back to the spin-2 character of the gravitational field, in contrast to the electromagnetism being a spin-1 field.[11] (See relativistic wave equations for more on "spin-1" and "spin-2" fields).

6.2.4 In Planck units

From comparison of GEM equations and Maxwell's equations it is obvious that $-1/(4\pi G)$ is the gravitational analog of vacuum permittivity ε_0. Adopting Planck units normalizes G, c and $1/(4\pi\varepsilon_0)$ to 1, thereby eliminating these constants from both sets of equations. The two sets of equations then become identical but for the minus sign preceding 4π in the GEM equations and a factor of four in Ampere's law. These minus signs stem from an essential difference between gravity and electromagnetism: electrostatic charges of identical sign repel each other, while masses attract each other. Hence the GEM equations are nearly Maxwell's equations with mass (or mass density) substituting for charge (or charge density), and $-G$ replacing the Coulomb force constant $1/(4\pi\varepsilon_0)$. 4π appears in both the GEM and Maxwell equations, because Planck units normalize G and $1/(4\pi\varepsilon_0)$ to 1, and not $4\pi G$ and $1/\varepsilon_0$.

6.3 Higher-order effects

Some higher-order gravitomagnetic effects can reproduce effects reminiscent of the interactions of more conventional polarized charges. For instance, if two wheels are spun on a common axis, the mutual gravitational attraction between the two wheels will be greater if they spin in opposite directions than in the same direction. This can be expressed as an attractive or repulsive gravitomagnetic component.

Gravitomagnetic arguments also predict that a flexible or fluid toroidal mass undergoing minor axis rotational acceleration (accelerating "smoke ring" rotation) will tend to pull matter through the throat (a case of rotational frame dragging, acting through the throat). In theory, this configuration might be used for accelerating objects (through the throat) without such objects experiencing any g-forces.[12]

Consider a toroidal mass with two degrees of rotation (both major axis and minor-axis spin, both turning inside out and revolving). This represents a "special case" in which gravitomagnetic effects generate a chiral corkscrew-like gravitational field around the object. The reaction forces to dragging at the inner and outer equators would normally be expected to be equal and opposite in magnitude and direction respectively in the simpler case involving only minor-axis spin. When *both* rotations are applied simultaneously, these two sets of reaction forces can be said to occur at different depths in a radial Coriolis field that extends across the rotating torus, making it more difficult to establish that cancellation is complete.

Modelling this complex behaviour as a curved spacetime problem has yet to be done and is believed to be very difficult.

6.4   Gravitomagnetic fields of astronomical objects

The formula for the gravitomagnetic field $\mathbf{B}_g$ near a rotating body can be derived from the GEM equations. It is given by:

$$\mathbf{B}_g = \frac{G}{2c^2} \frac{\mathbf{L} - 3(\mathbf{L} \cdot \mathbf{r}/r)\mathbf{r}/r}{r^3},$$

where $\mathbf{L}$ is the angular momentum of the body. At the equatorial plane, $\mathbf{r}$ and $\mathbf{L}$ are perpendicular, so their dot product vanishes, and this formula reduces to:

$$\mathbf{B}_g = \frac{G}{2c^2} \frac{\mathbf{L}}{r^3},$$

The magnitude of angular momentum of a homogeneous ball-shaped body is:

$$L = I_{\text{ball}}\omega = \frac{2mr^2}{5} \frac{2\pi}{T}$$

where:

- $I_{\text{ball}} = \frac{2mr^2}{5}$ is the moment of inertia of a ball-shaped body (see: list of moments of inertia);

- ω is the angular velocity;

- m is the mass;

- r is the radius;

- T is the rotational period.

6.4.1   Earth

Therefore, the magnitude of Earth's gravitomagnetic field at its equator is:

$$B_{\text{Earth g,}} = \frac{G}{5c^2} \frac{m}{r} \frac{2\pi}{T} = \frac{2\pi r g}{5c^2 T},$$

where $g = G\frac{m}{r^2}$ is Earth's gravity. The field direction coincides with the angular moment direction, i.e. north.

From this calculation it follows that Earth's equatorial gravitomagnetic field is about 1.012×10^{-14} Hz,[13] or 3.1×10^{-7} in units of standard gravity (9.81 m/s^2) divided by the speed of light. Such a field is extremely weak and requires extremely sensitive measurements to be detected. One experiment to measure such field was the Gravity Probe B mission.

6.4.2 Pulsar

If the preceding formula is used with the second fastest-spinning pulsar known, PSR J1748-2446ad (which rotates 716 times per second), assuming a radius of 16 km, and two solar masses, then

$$B_{\text{g}} = \frac{2\pi Gm}{5rc^2 T}$$

equals about 166 Hz. This would be easy to notice. However, the pulsar is spinning at a quarter of the speed of light at the equator, and its radius is only three times more than its Schwarzschild radius. When such fast motion and such strong gravitational fields exist in a system, the simplified approach of separating gravitomagnetic and gravitoelectric forces can be applied only as a very rough approximation.

6.5 Lack of invariance

While Maxwell's equations are invariant under Lorentz transformations, the GEM equations were not. The fact that ρ_g and j_g do not form a four-vector (instead they are merely a part of the stress–energy tensor) is the basis of this problem.

Although GEM may hold approximately in two different reference frames connected by a Lorentz boost, there is no way to calculate the GEM variables of one such frame from the GEM variables of the other, unlike the situation with the variables of electromagnetism. Indeed, their predictions (about what motion is free fall) will probably conflict with each other.

Note that the GEM equations are invariant under translations and spatial rotations, just not under boosts and more general curvilinear transformations. Maxwell's equations can be formulated in a way that makes them invariant under all of these coordinate transformations.

6.6 See also

- Linearized gravity

- Geodetic effect

- Gravitational radiation

- Gravity Probe B

- Frame-dragging

- Kaluza–Klein theory

- Speed of gravity#Electrodynamical analogies

6.7 References

[1] O. Heaviside (1893). "A gravitational and electromagnetic analogy". *The Electrician* **31**: 81–82.

[2] R. Penrose (1969). "Gravitational collapse: The role of general relativity". *Rivista de Nuovo Cimento*. Numero Speciale 1: 252–276. Bibcode:1969NCimR...1..252P.

[3] R.K. Williams (1995). "Extracting x rays, ϒ rays, and relativistic e⁻e⁺ pairs from supermassive Kerr black holes using the Penrose mechanism". *Physical Review* **51** (10): 5387–5427. Bibcode:1995PhRvD..51.5387W. doi:10.1103/PhysRevD.51.5387.

58   *CHAPTER 6. GRAVITOELECTROMAGNETISM*

[4] R.K. Williams (2004). "Collimated escaping vortical polar e^-e^+ jets intrinsically produced by rotating black holes and Penrose processes". *The Astrophysical Journal* **611** (2): 952–963. arXiv:astro-ph/0404135. Bibcode:2004ApJ...611..952W. doi:10.1086/422304.

[5] R.K. Williams (2005). "Gravitomagnetic field and Penrose scattering processes". *Annals of the New York Academy of Sciences* **1045**. pp. 232–245.

[6] R.K. Williams (2001). "Collimated energy–momentum extraction from rotating black holes in quasars and microquasars using the Penrose mechanism". *AIP Conference Proceedings* **586**. pp. 448–453. arXiv:astro-ph/0111161.

[7] Gravitation and Inertia, I. Ciufolini and J.A. Wheeler, Princeton Physics Series, 1995, ISBN 0-691-03323-4

[8] B. Mashhoon, F. Gronwald, H.I.M. Lichtenegger (1999). "Gravitomagnetism and the Clock Effect". *Lect.Notes Phys.* **562**: 83–108. arXiv:gr-qc/9912027. Bibcode:2001LNP...562...83M.

[9] S.J. Clark, R.W. Tucker (2000). "Gauge symmetry and gravito-electromagnetism". *Classical and Quantum Gravity* **17** (19): 4125–4157. arXiv:gr-qc/0003115. Bibcode:2000CQGra..17.4125C. doi:10.1088/0264-9381/17/19/311.

[10] B. Mashhoon (2008). "Gravitoelectromagnetism: A Brief Review". arXiv:gr-qc/0311030. Bibcode:2003gr.qc....11030M.

[11] B. Mashhoon (2000). "Gravitoelectromagnetism".arXiv:gr-qc/0011014.Bibcode:2001rfg..conf..121M.doi:10.1142/97898

[12] R.L. Forward (1963). "Guidelines to Antigravity". *American Journal of Physics* **31** (3): 166–170. Bibcode:1963AmJPh..31..166F. doi:10.1119/1.1969340.

[13] http://www.google.com/search?q=2*pi*radius+of+Earth*earth+gravity%2F(5*c\char"005E\relax{}2*day)

6.8 Further reading

6.8.1 Books

- M. P. Hobson, G. P. Efstathiou, A. N. Lasenby (2006). *General Relativity: An Introduction for Physicists*. Cambridge University Press. p. 490–491. ISBN 9780521829519.

- L. H. Ryder (2009).*Introduction to General Relativity*. Cambridge University Press. p. 200–207.ISBN97805218

- J. B. Hartle (2002). *Gravity: An Introduction to Einstein's General Relativity*. Addison-Wesley. p. 296, 303. ISBN 9780805386622.

- S. Carroll (2003). *Spacetime and Geometry: An Introduction to General Relativity*. Addison-Wesley. p. 281. ISBN 9780805387322.

- J.A. Wheeler (1990). "Gravity's next prize: Gravitomagnetism". *A journey into gravity and spacetime*. Scientific American Library. pp. 232–233. ISBN 0-7167-5016-3.

- L. Iorio (ed.) (2007). *Measuring Gravitomagnetism: A Challenging Enterprise*. Nova. ISBN 1-60021-002-3.

- O.D. Jefimenko (1992). *Causality, electromagnetic induction, and gravitation : a different approach to the theory of electromagnetic and gravitational fields*. Electret Scientific. ISBN 0-917406-09-5.

- O.D. Jefimenko (2006). *Gravitation and Cogravitation*. Electret Scientific. ISBN 0-917406-15-X.

6.8.2 Papers

- S.J. Clark, R.W. Tucker (2000). "Gauge symmetry and gravito-electromagnetism". *Classical and Quantum Gravity* **17** (19): 4125–4157. arXiv:gr-qc/0003115. Bibcode:2000CQGra..17.4125C. doi:10.1088/0264-9381/17/19/311.

- R.L. Forward (1963). "Guidelines to Antigravity".*American Journal of Physics***31**(3): 166–170.Bibcode:1963 doi:10.1119/1.1969340.

- R.T. Jantzen, P. Carini, D. Bini (1992). "The Many Faces of Gravitoelectromagnetism". *Annals of Physics* **215**: 1–50. arXiv:gr-qc/0106043. Bibcode:1992AnPhy.215....1J. doi:10.1016/0003-4916(92)90297-Y.

- B. Mashhoon (2000). "Gravitoelectromagnetism". arXiv:gr-qc/0011014 [gr-qc].

- B. Mashhoon (2003). "Gravitoelectromagnetism: a Brief Review". arXiv:gr-qc/0311030 [gr-qc]. in L. Iorio (ed.) (2007). *Measuring Gravitomagnetism: A Challenging Enterprise*. Nova. pp. 29–39. ISBN 1-60021-002-3.

- M. Tajmar, C.J. de Matos (2001). "Gravitomagnetic Barnett Effect". *Indian Journal of Physics B* **75**: 459–461. arXiv:gr-qc/0012091. Bibcode:2000gr.qc....12091D.

- L. Filipe Costa, Carlos A. R. Herdeiro (2007). "A gravito-electromagnetic analogy based on tidal tensors". *Physical Review D* **78** (2). arXiv:gr-qc/0612140. Bibcode:2008PhRvD..78b4021C. doi:10.1103/PhysRevD.78.024021.

- Antoine Acke (2013). "Gravito-electromagnetism explained by the theory of informatons". *Hadronic Journal* **36** (4).

6.9 External links

- Gravity Probe B: Testing Einstein's Universe

- Gyroscopic Superconducting Gravitomagnetic Effects news on tentative result of European Space Agency (esa) research

- In Search of Gravitomagnetism, NASA, 20 April 2004.

- Gravitomagnetic London Moment – New test of General Relativity?

- Measurement of Gravitomagnetic and Acceleration Fields Around Rotating Superconductors M. Tajmar, et al., 17 October 2006.

- Test of the Lense–Thirring effect with the MGS Mars probe, *New Scientist*, January 2007.

Chapter 7

Gravitational-wave observatory

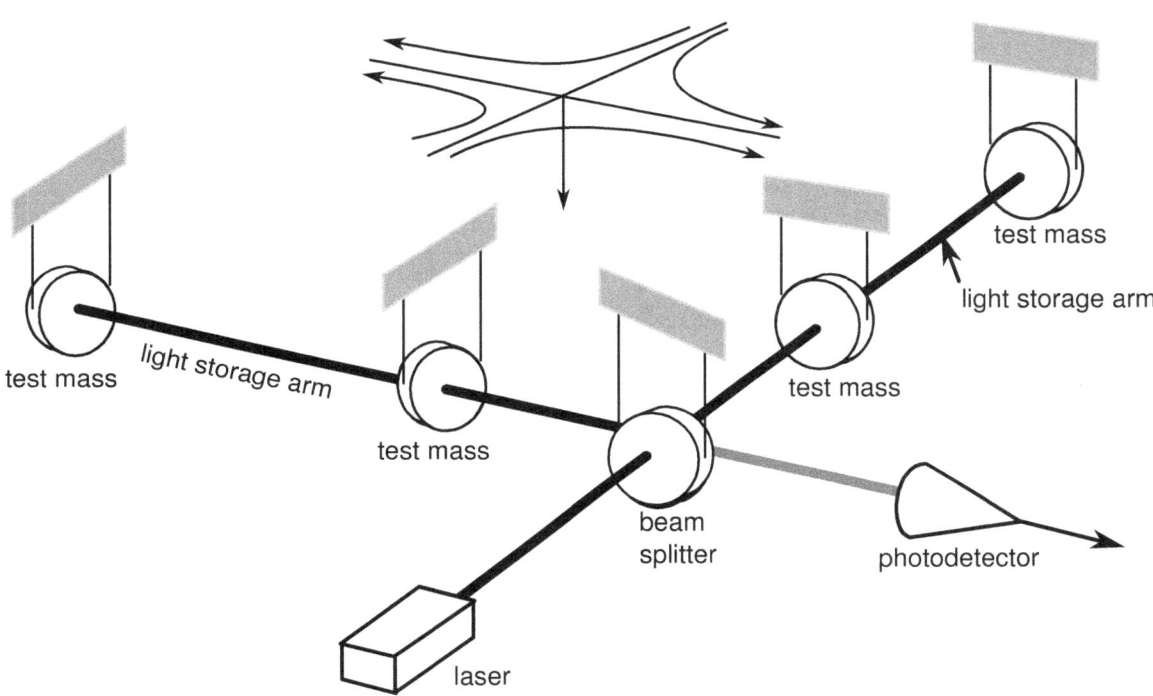

A schematic diagram of a laser interferometer.

A **gravitational-wave observatory** (or **gravitational-wave detector**) is any device designed to measure gravitational waves, tiny distortions of spacetime that were first predicted by Einstein in 1916.[1] Gravitational waves are perturbations in the curvature of spacetime caused by accelerated masses. The existence of gravitational radiation is a specific prediction of general relativity, but is a feature of all theories of gravity that obey special relativity.[2] Since the 1960s, gravitational-wave detectors have been built and constantly improved. The present-day generation of resonant mass antennas and laser interferometers has reached the necessary sensitivity to detect gravitational waves from sources in the Milky Way. Gravitational-wave observatories are the primary tool of gravitational-wave astronomy.

As of late 2015, no direct detection of gravitational waves had been accomplished. However, a number of experiments had provided evidence that gravitational waves did exist, notably the observation of binary pulsars, the orbits of which evolve precisely matching the predictions of energy loss through general relativistic gravitational-wave emission. The 1993 Nobel Prize in Physics was awarded for this work.[3]

On 17 March 2014, astronomers at the Harvard-Smithsonian Center for Astrophysics announced the apparent detection of the imprint gravitational waves in the cosmic microwave background, which, if confirmed, would provide strong evidence for inflation and the Big Bang.[4][5][6][7] However, on 19 June 2014, lowered confidence in confirming the findings was reported;[8][9][10] and on 19 September 2014, even more lowered confidence.[11][12] Finally, on January 30, 2015, the European Space Agency announced that the signal can be entirely attributed to dust in the Milky Way.[13]

7.1 Complications

The direct detection of gravitational waves is complicated by the extraordinarily small effect the waves would produce on a detector. The amplitude of a spherical wave will fall off as the inverse of the distance from the source. Thus, even waves from extreme systems like merging binary black holes die out to very small amplitude by the time they reach the Earth. Astrophysicists expect that some gravitational waves passing the Earth may be as large as $h \approx 10^{-18}$, but generally no bigger.

7.2 Weber bars

A simple device to detect the expected wave motion is called a Weber bar – a large, solid bar of metal isolated from outside vibrations. This type of instrument was the first type of gravitational-wave detector. Strains in space due to an incident gravitational wave excite the bar's resonant frequency and could thus be amplified to detectable levels. Conceivably, a nearby supernova might be strong enough to be seen without resonant amplification. Modern forms of the Weber bar are still operated, cryogenically cooled, with superconducting quantum interference devices to detect vibration (see for example, ALLEGRO). Weber bars are not sensitive enough to detect anything but extremely powerful gravitational waves.[14]

MiniGRAIL is a spherical gravitational-wave antenna using this principle. It is based at Leiden University, consisting of an exactingly machined 1150 kg sphere cryogenically cooled to 20 mK.[15] The spherical configuration allows for equal sensitivity in all directions, and is somewhat experimentally simpler than larger linear devices requiring high vacuum. Events are detected by measuring deformation of the detector sphere. MiniGRAIL is highly sensitive in the 2–4 kHz range, suitable for detecting gravitational waves from rotating neutron star instabilities or small black hole mergers.[16]

AURIGA is an ultracryogenic resonant bar gravitational wave detector based at INFN in Italy. It is based on a cylindrical bar detector. The AURIGA and LIGO teams have collaborated in joint observations.[17]

7.3 Interferometers

A more sensitive detector uses laser interferometry to measure gravitational-wave induced motion between separated 'free' masses.[18] This allows the masses to be separated by large distances (increasing the signal size); a further advantage is that it is sensitive to a wide range of frequencies (not just those near a resonance as is the case for Weber bars). Ground-based interferometers are now operational. Currently, the most sensitive is LIGO – the Laser Interferometer Gravitational Wave Observatory. LIGO has three detectors: one in Livingston, Louisiana; the other two (in the same vacuum tubes) at the Hanford site in Richland, Washington. Each consists of two light storage arms which are 2 to 4 kilometers in length. These are at 90 degree angles to each other, with the light passing through 1m diameter vacuum tubes running the entire 4 kilometers. A passing gravitational wave will slightly stretch one arm as it shortens the other. This is precisely the motion to which an interferometer is most sensitive.

Even with such long arms, the strongest gravitational waves will only change the distance between the ends of the arms by at most roughly 10^{-18} meters. LIGO should be able to detect gravitational waves as small as $h \approx 5 \times 10^{-22}$. Upgrades to LIGO and other detectors such as VIRGO, GEO 600, and TAMA 300 should increase the sensitivity still further; the next generation of instruments (Advanced LIGO and Advanced Virgo) will be more than ten times more sensitive. Another highly sensitive interferometer (LCGT) is currently in the design phase. A key point is that a ten-times increase in sensitivity (radius of "reach") increases the volume of space accessible to the instrument by one thousand. This increases

Atomic interferometry.

the rate at which detectable signals should be seen from one per tens of years of observation, to tens per year.

Interferometric detectors are limited at high frequencies by shot noise, which occurs because the lasers produce photons randomly; one analogy is to rainfall – the rate of rainfall, like the laser intensity, is measurable, but the raindrops, like photons, fall at random times, causing fluctuations around the average value. This leads to noise at the output of the detector, much like radio static. In addition, for sufficiently high laser power, the random momentum transferred to the test masses by the laser photons shakes the mirrors, masking signals at low frequencies. Thermal noise (e.g., Brownian motion) is another limit to sensitivity. In addition to these "stationary" (constant) noise sources, all ground-based detectors are also limited at low frequencies by seismic noise and other forms of environmental vibration, and other "non-stationary" noise sources; creaks in mechanical structures, lightning or other large electrical disturbances, etc. may also create noise masking an event or may even imitate an event. All these must be taken into account and excluded by analysis before a detection may be considered a true gravitational-wave event.

Space-based interferometers, such as LISA and DECIGO, are also being developed. LISA's design calls for three test masses forming an equilateral triangle, with lasers from each spacecraft to each other spacecraft forming two independent interferometers. LISA is planned to occupy a solar orbit trailing the Earth, with each arm of the triangle being five million kilometers. This puts the detector in an excellent vacuum far from Earth-based sources of noise, though it will still be susceptible to shot noise, as well as artifacts caused by cosmic rays and solar wind.

An *atomic gravitational wave interferometric sensor* (AGIS) is a novel detection scheme to detect gravitational waves, proposed by S. Dimopoulos et al. in 2008.[19][20]

7.4   High frequency detectors

There are currently two detectors focusing on detections at the higher end of the gravitational-wave spectrum (10^{-7} to 10^5 Hz): one at University of Birmingham, England, and the other at INFN Genoa, Italy. A third is under development at Chongqing University, China. The Birmingham detector measures changes in the polarization state of a microwave beam circulating in a closed loop about one meter across. Two have been fabricated and they are currently expected to be sensitive to periodic spacetime strains of $h \sim 2 \times 10^{-13}/\sqrt{Hz}$, given as an amplitude spectral density. The INFN Genoa detector is a resonant antenna consisting of two coupled spherical superconducting harmonic oscillators a few centimeters in diameter. The oscillators are designed to have (when uncoupled) almost equal resonant frequencies. The system is currently expected to have a sensitivity to periodic spacetime strains of $h \sim 2 \times 10^{-17}/\sqrt{Hz}$, with an expectation to reach a sensitivity of $h \sim 2 \times 10^{-20}/\sqrt{Hz}$. The Chongqing University detector is planned to detect relic high-frequency gravitational waves with the predicted typical parameters $?_g \sim 10^{10}$ Hz (10 GHz) and $h \sim 10^{-30}$–10^{-31}.

7.5 Pulsar timing arrays

A different approach to detecting gravitational waves is used by pulsar timing arrays, such as the European Pulsar Timing Array,[21] the North American Nanohertz Observatory for Gravitational Waves,[22] and the Parkes Pulsar Timing Array.[23] These projects propose to detect gravitational waves by looking at the effect these waves have on the incoming signals from an array of 20–50 well-known millisecond pulsars. As a gravitational wave passing through the Earth contracts space in one direction and expands space in another, the times of arrival of pulsar signals from those directions are shifted correspondingly. By studying a fixed set of pulsars across the sky, these arrays should be able to detect gravitational waves in the nanohertz range. Such signals are expected to be emitted by pairs of merging supermassive black holes.[24]

7.6 Einstein@Home

Main article: Einstein@Home

In some sense, the easiest signals to detect should be constant sources. Supernovae and neutron star or black hole mergers should have larger amplitudes and be more interesting, but the waves generated will be more complicated. The waves given off by a spinning, bumpy neutron star would be "monochromatic" – like a pure tone in acoustics. It would not change very much in amplitude or frequency.

The Einstein@Home project is a distributed computing project similar to SETI@home intended to detect this type of simple gravitational wave. By taking data from LIGO and GEO, and sending it out in little pieces to thousands of volunteers for parallel analysis on their home computers, Einstein@Home can sift through the data far more quickly than would be possible otherwise.[25]

7.7 Specific operational and planned gravitational-wave detectors

- CLIO

- GEO 600

- KAGRA

- LIGO

- MiniGrail

- Pulsar timing array

- TAMA 300

- Virgo interferometer

- eLISA

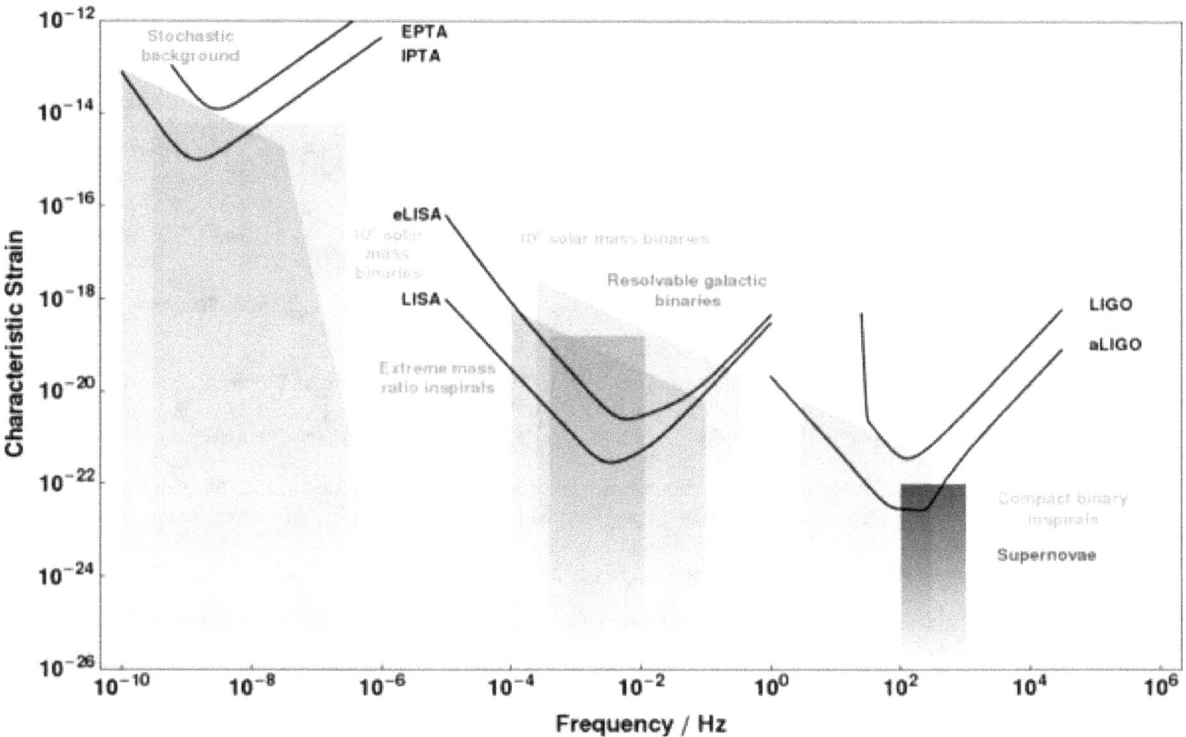

Noise curves for a selection of detectors as a function of frequency. The characteristic strain of potential astrophysical sources are also shown. To be detectable the characteristic strain of a signal must be above the noise curve.[26]

7.8   References

[1] Clark, Stuart (17 March 2014). "What are gravitational waves?". *The Guardian*. Retrieved 22 May 2014.

[2] Schutz, Bernard F. (1984). "Gravitational waves on the back of an envelope". *American Journal of Physics* **52** (5): 412. Bibcode:1984AmJPh..52..412S. doi:10.1119/1.13627.

[3] "Press Release: The Nobel Prize in Physics 1993". Nobel Prize. 13 October 1993. Retrieved 6 May 2014.

[4] Staff (17 March 2014). "BICEP2 2014 Results Release". *National Science Foundation*. Retrieved 18 March 2014.

[5] Clavin, Whitney (17 March 2014). "NASA Technology Views Birth of the Universe". *NASA*. Retrieved 17 March 2014.

[6] Overbye, Dennis (17 March 2014). "Detection of Waves in Space Buttresses Landmark Theory of Big Bang". *New York Times*. Retrieved 17 March 2014.

[7] Overbye, Dennis (24 March 2014). "Ripples From the Big Bang". *New York Times*. Retrieved 24 March 2014.

[8] Overbye, Dennis (19 June 2014). "Astronomers Hedge on Big Bang Detection Claim". *New York Times*. Retrieved 20 June 2014.

[9] Amos, Jonathan (19 June 2014). "Cosmic inflation: Confidence lowered for Big Bang signal". *BBC News*. Retrieved 20 June 2014.

[10] Ade, P.A.R. (BICEP2 Collaboration); et al. (19 June 2014). "Detection of B-Mode Polarization at Degree Angular Scales by BICEP2". *Physical Review Letters* **112** (24): 241101. arXiv:1403.3985. Bibcode:2014PhRvL.112x1101A. doi:10.1103/PhysRev Lett.112.241101.PMID24996078.

[11] Planck Collaboration Team (19 September 2014). "Planck intermediate results. XXX. The angular power spectrum of polarized dust emission at intermediate and high Galactic latitudes". *ArXiv*. arXiv:1409.5738. Bibcode:2014arXiv1409.5738P. Retrieved 22 September 2014.

[12] Overbye, Dennis (22 September 2014). "Study Confirms Criticism of Big Bang Finding". *New York Times*. Retrieved 22 September 2014.

[13] Cowen, Ron (2015-01-30). "Gravitational waves discovery now officially dead". *nature*. doi:10.1038/nature.2015.16830.

[14] For a review of early experiments using Weber bars, see Levine, J. (April 2004). "Early Gravity-Wave Detection Experiments, 1960-1975". *Physics in Perspective (Birkhäuser Basel)* **6** (1): 42–75. Bibcode:2004PhP.....6...42L. doi:10.1007/s00016-003-0179-6.

[15] Gravitational Radiation Antenna In Leiden

[16] de Waard, Arlette; Luciano Gottardi; Giorgio Frossati (2000). "Marcel Grossmann meeting on General Relativity" (PDF). Rome, Italy. |contribution= ignored (help)

[17] AURIGA Collaboration; LIGO Scientific Collaboration; Baggio; Cerdonio, M; De Rosa, M; Falferi, P; Fattori, S; Fortini, P; et al. (2008). "A Joint Search for Gravitational Wave Bursts with AURIGA and LIGO". *Classical and Quantum Gravity* **25** (9): 095004. arXiv:0710.0497. Bibcode:2008CQGra..25i5004B. doi:10.1088/0264-9381/25/9/095004.

[18] The idea of using laser interferometry for gravitational-wave detection was first mentioned by Gerstenstein and Pustovoit 1963 Sov. Phys.–JETP 16 433. Weber mentioned it in an unpublished laboratory notebook. Rainer Weiss first described in detail a practical solution with an analysis of realistic limitations to the technique in R. Weiss (1972). "Electromagnetically Coupled Broadband Gravitational Antenna". Quarterly Progress Report, Research Laboratory of Electronics, MIT 105: 54.

[19] Bender, Peter L. "Comment on "Atomic gravitational wave interferometric sensor"". *Physical Review D* **84** (2). Bibcode:2011 doi:10.1103/PhysRevD.84.028101. PhRvD..84b8101B.

[20] Johnson, David Marvin Slaughter (2011). "AGIS-LEO". *Long Baseline Atom Interferometry*. Stanford University. pp. 41–98.

[21] Janssen, G. H.; Stappers, B. W.; Kramer, M.; Purver, M.; Jessner, A.; Cognard, I.; Bassa, C.; Wang, Z.; Cumming, A.; Kaspi, V. M. (2008). "European Pulsar Timing Array". *AIP Conference Proceedings* **983**: 633–635. doi:10.1063/1.2900317.

[22] North American Nanohertz Observatory for Gravitational Waves (NANOGrav) homepage

[23] Parkes Pulsar Timing Array homepage

[24] Hobbs, G. B.; Bailes, M.; Bhat, N. D. R.; Burke-Spolaor, S.; Champion, D. J.; Coles, W.; Hotan, A.; Jenet, F.; et al. (2008). "Gravitational wave detection using pulsars: status of the Parkes Pulsar Timing Array project". arXiv:0812.2721 [astro-ph].

[25] Einstein@Home

[26] Moore, Christopher; Cole, Robert; Berry, Christopher (19 July 2013). "Gravitational Wave Detectors and Sources". Retrieved 17 April 2014.

Chapter 8

LIGO

For the Latvian holiday Ligo, see Jāņi.

The **Laser Interferometer Gravitational-Wave Observatory (LIGO)** is a large-scale physics experiment aiming to directly detect gravitational waves. Cofounded in 1992 by Kip Thorne and Ronald Drever of Caltech and Rainer Weiss of MIT, LIGO is a joint project between scientists at MIT, Caltech, and many other colleges and universities. It is sponsored by the National Science Foundation (NSF). At the cost of $620 million,[1] it is the largest and most ambitious project ever funded by the NSF.[2][3]

Initial LIGO operations between 2002 and 2010 did not detect any gravitational waves. This was followed by a multi-year shutdown while the detectors were replaced by much improved "Advanced LIGO" versions. As of February 2015, two such advanced detectors (one in Livingston, Louisiana and the other in Hanford, Washington) have been brought into engineering mode.[4] On September 18, 2015, Advanced LIGO became fully operational and began formal science operations at twice the sensitivity of the initial LIGO interferometers.[5]

8.1 Mission

LIGO's mission is to directly observe gravitational waves of cosmic origin. These waves were first predicted by Einstein's general theory of relativity in 1916, when the technology necessary for their detection did not yet exist. Their existence was indirectly confirmed when observations of the binary pulsar PSR 1913+16 in 1974 showed an orbital decay which matched Einstein's predictions of energy loss by gravitational radiation. The Nobel Prize in Physics 1993 was awarded to Hulse and Taylor for this discovery.[7]

Direct detection of gravitational waves has long been sought. Their discovery would launch a new branch of astronomy to complement electromagnetic telescopes and neutrino observatories. Joseph Weber pioneered the effort to detect gravitational waves in the 1960s through his work on resonant mass bar detectors. Bar detectors continue to be used at six sites worldwide. By the 1970s, scientists including Rainer Weiss realized the applicability of laser interferometry to gravitational wave measurements. Robert Forward operated an interferometric detector at Hughes in the early 1970s.[8]

In fact as early as the 1960s, and perhaps before that, there were papers published on wave resonance of light and gravitational waves.[9] Work was published in 1971 on methods to exploit this resonance for the detection of high-frequency gravitational waves. In 1962, M. E. Gertsenshtein and V. I. Pustovoit published the very first paper describing the principles for using interferometers for the detection of very long wavelength gravitational waves.[10] The authors argued that by using interferometers the sensitivity can be 10^7–10^{10} times better than by using electromechanical experiments. Later, in 1965, Braginsky, extensively discussed gravitational-wave sources and their possible detection. He pointed out the 1962 paper and mentioned the possibility of detecting gravitational waves if the interferometric technology and measuring techniques improved.

In August 2002, LIGO began its search for cosmic gravitational waves. Measurable emissions of gravitational waves

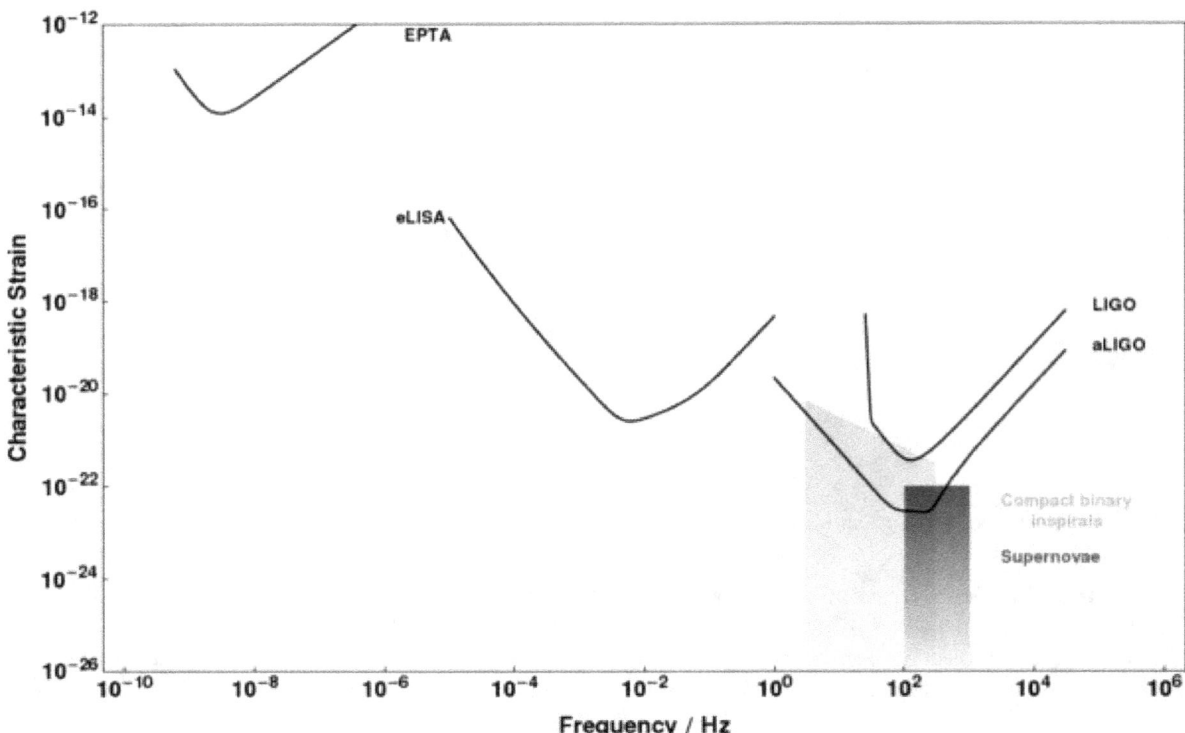

Detector noise curves for Initial and Advanced LIGO as a function of frequency. They lie above the bands for space-borne detectors like the evolved Laser Interferometer Space Antenna (eLISA) and pulsar timing arrays such as the European Pulsar Timing Array (EPTA). The characteristic strain of potential astrophysical sources are also shown. To be detectable the characteristic strain of a signal must be above the noise curve.[6]

are expected from binary systems (collisions and coalescences of neutron stars or black holes), supernova explosions of massive stars (which form neutron stars and black holes), accreting neutron stars, rotations of neutron stars with deformed crusts, and the remnants of gravitational radiation created by the birth of the universe. The observatory may, in theory, also observe more exotic hypothetical phenomena, such as gravitational waves caused by oscillating cosmic strings or colliding domain walls. Since the early 1990s, physicists have thought that technology has evolved to the point where detection of gravitational waves—of significant astrophysical interest—is now possible.[11]

8.2 Observatories

LIGO operates two gravitational wave observatories in unison: the LIGO Livingston Observatory(30°33′46.42″N90°46′27. 27″W/30.5628944°N90.7742417°W) in Livingston, Louisiana, and the LIGO Hanford Observatory, on the DOE Hanford Site(46°27′18.52″N119°24′27.56″W/46.4551444°N119.4076556°W), located near Richland, Washington. These sites are separated by 3,002 kilometers (1,865miles).
Since gravitational waves are expected to travel at the speed of light,this distance corresponds to a difference in gravitational wave arrival times of up to ten milliseconds. Through the use of triangulation, the difference in arrival times can determine the source of the wave in the sky.

Each observatory supports an L-shaped ultra high vacuum system, measuring 4 kilometers (2.5 miles) on each side. Up to five interferometers can be set up in each vacuum system.

The LIGO Livingston Observatory houses one laser interferometer in the primary configuration. This interferometer was successfully upgraded in 2004 with an active vibration isolation system based on hydraulic actuators providing a factor of 10 isolation in the 0.1 – 5 Hz band. Seismic vibration in this band is chiefly due to microseismic waves and anthropogenic sources (traffic, logging, etc.).

Northern leg (x-arm) of LIGO interferometer on Hanford Reservation

The LIGO Hanford Observatory houses one interferometer, almost identical to the one at the Livingston Observatory. During the Initial and Enhanced LIGO eras, a half-length interferometer was operated in parallel with the main interferometer. For this 2 km interferometer, the Fabry–Pérot arm cavities had the same optical finesse, and thus half the storage time, as the 4 km interferometers. With half the storage time, the theoretical strain sensitivity was as good as the full length interferometers above 200 Hz but only half as good at low frequencies. During the same era, Hanford retained its original passive seismic isolation system due to limited geologic activity in Southeastern Washington.

8.3   Operation

The primary interferometer at each site consists of mirrors suspended at each of the corners of the L; it is known as a power-recycled Michelson interferometer with Gires–Tournois etalon arms. A pre-stabilized laser emits a beam of up to 200 Watts that passes through an optical mode cleaner before reaching a beam splitter at the vertex of the L. There the beam splits into two paths, one for each arm of the L; each arm contains Fabry–Pérot cavities that store the beams and increase the effective path length.

When a gravitational wave passes through the interferometer, the space-time in the local area is altered. Depending on the source of the wave and its polarization, this results in an effective change in length of one or both of the cavities. The effective length change between the beams will cause the light currently in the cavity to become very slightly out of phase with the incoming light. The cavity will therefore periodically get very slightly out of resonance and the beams which are tuned to destructively interfere at the detector, will have a very slight periodically varying detuning. This results in a measurable signal. Note that the effective length change and the resulting phase change are a subtle tidal effect that must be carefully computed because the light waves are affected by the gravitational wave just as much as the beams themselves.[12]

After an equivalent of approximately 75 trips down the 4 km length to the far mirrors and back again, the two separate beams leave the arms and recombine at the beam splitter. The beams returning from two arms are kept out of phase so that when the arms are both in resonance (as when there is no gravitational wave passing through), their light waves subtract, and no light should arrive at the photodiode. When a gravitational wave passes through the interferometer, the distances along the arms of the interferometer are shortened and lengthened, causing the beams to become slightly less out of phase, so some light arrives at the photodiode, indicating a signal. Light that does not contain a signal is returned to the interferometer using a power recycling mirror, thus increasing the power of the light in the arms. In actual operation, noise sources can cause movement in the optics which produces similar effects to real gravitational wave signals; a great deal of the art and complexity in the instrument is in finding ways to reduce these spurious motions of the mirrors. Observers compare signals from both sites to reduce the effects of noise.

8.4   Observations

Western leg of LIGO interferometer on Hanford Reservation

Based on current models of astronomical events, and the predictions of the general theory of relativity, gravitational waves that originate tens of millions of light years from Earth are expected to distort the 4 kilometer mirror spacing by about 10^{-18} m, less than one-thousandth the charge diameter of a proton. Equivalently, this is a relative change in distance of approximately one part in 10^{21}. A typical event which might cause a detection event would be the late stage inspiral and merger of two 10 solar mass black holes, not necessarily located in the Milky Way galaxy, which is expected to result in a very specific sequence of signals often summarized by the slogan *chirp, burst, quasi-normal mode ringing, exponential decay.*

In their fourth Science Run at the end of 2004, the LIGO detectors demonstrated sensitivities in measuring these displacements to within a factor of 2 of their design.

During LIGO's fifth Science Run in November 2005, sensitivity reached the primary design specification of a detectable strain of one part in 10^{21} over a 100 Hz bandwidth. The baseline inspiral of two roughly solar-mass neutron stars is typically expected to be observable if it occurs within about 8,000,000 parsecs (26,000,000 ly), or the vicinity of our Local Group of galaxies, averaged over all directions and polarizations. Also at this time, LIGO and GEO 600 (the German-UK interferometric detector) began a joint science run, during which they collected data for several months. Virgo (the French-Italian interferometric detector) joined in May 2007. The fifth science run ended in 2007. After extensive analysis data from this run did not uncover any unambiguous detection events.

In February 2007, GRB 070201, a short gamma-ray burst, arrived at Earth from the direction of the Andromeda Galaxy, a nearby galaxy. The prevailing explanation of most short gamma-ray bursts is the merger of a neutron star with either a neutron star or black hole. LIGO reported a non-detection for GRB 070201, ruling out a merger at the distance of Andromeda with high confidence. Such a constraint is predicated on LIGO eventually demonstrating a direct detection of gravitational waves.[13]

8.4.1 Enhanced LIGO

After the completion of Science Run 5, initial LIGO was upgraded with certain Advanced LIGO technologies that resulted in an improved-performance configuration dubbed Enhanced LIGO.[14] Some of the improvements in Enhanced LIGO included:

- Increased laser power

- Homodyne detection

- Output mode cleaner

- In-vacuum readout hardware

Science Run 6 (S6) began in July 2009 with the enhanced configurations on the 4 km detectors.[15] It concluded in October 2010, and the disassembling of the original detectors began. A five-year-long effort to install and commission the Advanced LIGO detectors was completed in September 2015.

8.5 Future

8.5.1 Advanced LIGO

The LIGO Laboratory, funded by the National Science Foundation with contributions from the GEO 600 Collaboration and ANU and Adelaide Universities in Australia, and with participation by the LIGO Scientific Collaboration, has installed the new Advanced LIGO detectors in the LIGO Observatory infrastructures. This new detector is designed to improve the sensitivity of initial LIGO by more than a factor of 10 once fully commissioned. The LIGO Laboratory started the first Observing Run 'O1' with the Advanced LIGO detectors in September 2015 at a sensitivity roughly 4 times greater than Initial LIGO for some classes of sources (e.g., neutron-star binaries), and a much greater sensitivity for larger systems with their peak radiation at lower audio frequencies.[16] Further Observing runs will be interleaved with commissioning efforts to further improve the sensitivity.

8.5.2 LIGO-India

LIGO-India is a proposed collaborative project between the LIGO Laboratory and the Indian Initiative in Gravitational Observations (IndIGO) to create a world-class gravitational-wave detector in India. The LIGO Laboratory, with permission from the U.S. National Science Foundation and Advanced LIGO partners from the U.K, Germany and Australia, has offered to provide all of the designs and hardware for one of the two planned Advanced LIGO detectors to be installed, commissioned, and operated by an Indian team of scientists in a facility to be built in India.

The expansion of worldwide activities in gravitational-wave detection to produce an effective global network has been a goal of LIGO for many years. In 2010, a developmental roadmap[17] issued by the Gravitational Wave International Committee (GWIC) recommended that an expansion of the global array of interferometric detectors be pursued as a highest priority. Such a network would afford astrophysicists with more robust search capabilities and higher scientific yields. The current agreement between the LIGO Scientific Collaboration and the Virgo collaboration links three comparable sensitivity detectors and forms the core of this international network. A fourth site not in the plane formed by the present three and distant from them all greatly improves source localization ability. Studies indicate that the localization of sources by a network that includes a detector in India would provide significant improvements.[18][19] Improvements in localization averages are predicted to be approximately an order of magnitude, with substantially larger improvements in certain regions of the sky.

The NSF was willing to permit this relocation, and its consequent schedule delays, as long as it did not increase the LIGO budget. Thus, all costs required to build a laboratory equivalent to the LIGO sites to house the detector would have to be borne by the host country.[20] The first potential distant location was at AIGO in Western Australia,[21] however the Australian government was unwilling to commit funding by the 1 October 2011 deadline.

A location in India was discussed at a Joint Commission meeting between India and the US in June 2012.[22] In parallel, the proposal was evaluated by LIGO's funding agency, the NSF. As the basis of the LIGO-India project entails the transfer of one of LIGO's detectors to India, the plan would affect work and scheduling on the Advanced LIGO upgrades already underway. In August 2012, the U.S. National Science Board approved the LIGO Laboratory's request to modify the scope of Advanced LIGO by not installing the Hanford "H2" interferometer, and to prepare it instead for storage in anticipation of sending it to LIGO-India.[23] In India, the project has been presented to the Department of Atomic Energy and the Department of Science and Technology for approval and funding. Final approval has not been granted yet.

8.6 See also

- Einstein Telescope, for a European third-generation gravitational wave detector.

- Einstein@Home, for a volunteer distributed computing program one can download in order to help the LIGO/GEO teams analyze their data.

- Fermilab Holometer

- GEO 600, for a gravitational wave detector located in Hannover, Germany.

- List of laser articles

- Tests of general relativity

- Virgo interferometer, an interferometer similar to LIGO, located close to Pisa, Italy.

8.7 Notes

[1] Zhang, Sarah (15 September 2015). "The Long Search for Elusive Ripples in Spacetime".

[2] Larger physics projects in the United States, such as Fermilab, have traditionally been funded by the Department of Energy.

[3] LIGO Fact Sheet at NSF

[4] "LIGO Hanford's H1 Achieves Two-Hour Full Lock". February 2015.

[5] Amos, Jonathan (19 September 2015). "Advanced Ligo: Labs 'open their ears' to the cosmos". *BBC News*. Retrieved 19 September 2015.

[6] Moore, Christopher; Cole, Robert; Berry, Christopher (19 July 2013). "Gravitational Wave Detectors and Sources". Retrieved 20 April 2014.

[7] "The Nobel Prize in Physics 1993: Russell A. Hulse, Joseph H. Taylor Jr.". *nobelprize.org*.

[8] California Institute of Technology announces death of Robert L Forward September 22, 2002

[9] V. B. Braginsky, L. P. Grishchuck, A. G. Doroshkevich, M .B. Mensky, I.D.Novikov, M. V. Sazhin and Y. B. Zeldovisch

[10] Gertsenshtein, M. E.; Pustovoit, V. I. (August 1962). "On the detection of low frequency gravitational waves". *JETP* **43**: 605–607.

[11] "Astrophysical Sources of Gravitational Radiation".

[12] Thorne, Kip (2004). "Chapter 26.5: The Detection of Gravitational Waves (in "Applications of Classical Physics chapter 26: Gravitational Waves and Experimental Tests of General Relativity", Caltech lecture notes)" (PDF). Retrieved 2010-08-02.

[13] "LIGO Sheds Light on Cosmic Event". 2007-12-20. Retrieved 2007-12-21.

[14] Adhikari, Fritschel, and Waldman. LIGO technical document LIGO-T060156-01-I. July 17th, 2006.

[15] Firm Date Set for Start of S6, by Dave Beckett, 6/15/2009, LIGO Laboratory News

[16] LIGO Scientific Collaboration. "Advanced Ligo". *http://arxiv.org/abs/1411.4547*.

[17] GWIC Developmental Roadmap p. 97

[18] Fairhurst, Stephen (28 Sep 2012), *Improved Source Localization with LIGO India*, LIGO document P1200054-v6

[19] Schutz, Bernard F. (25 Apr 2011), *Networks of Gravitational Wave Detectors and Three Figures of Merit*, arXiv:1102.5421, Bibcode:2011CQGra..2815023S, doi:10.1088/0264-9381/28/12/125023

[20] Cho, Adrian (27 August 2010), "U.S. Physicists Eye Australia for New Site of Gravitational-Wave Detector" (PDF), *Science* **329** (5995): 1003, Bibcode:2010Sci...329.1003C, doi:10.1126/science.329.5995.1003

[21] Finn, Sam; Fritschel, Peter; Klimenko, Sergey; Raab, Fred; Sathyaprakash, B.; Saulson, Peter; Weiss, Rainer (13 May 2010), *Report of the Committee to Compare the Scientific Cases for AHLV and HHLV*, LIGO document T1000251-v1

[22] U.S.-India Bilateral Cooperation on Science and Technology meeting fact sheet – dated June 13, 2012.

[23] Memorandum to Members and Consultants of the National Science Board – dated August 24, 2012

8.8   References

- Kip Thorne, ITP & Caltech. *Spacetime Warps and the Quantum: A Glimpse of the Future.* Lecture slides and audio

- Rainer Weiss, *Electromagnetically coupled broad-band gravitational wave antenna*, MIT RLE QPR 1972

- On the detection of low frequency gravitational waves, M.E.Gertsenshtein and V.I.Pustovoit – JETP Vol.43 p. 605-607 (August 1962) Note: This is the first paper proposing the use of interferometers for the detection of gravitational waves.

- Wave resonance of light and gravitational waves – M.E.Gertsenshtein – JETP Vol.41 p. 113-114 (July 1961)

- Gravitational electromagnetic resonance, V.B.Braginskii, M.B.Mensky – GR.G. Vol.3 No.4 p. 401-402 (1972)

- Gravitational radiation and the prospect of its experimental discovery, V.B.Braginsky – Soviet Physics Vol.86 p. 433-446 (July 1965)

- On the electromagnetic detection of gravitational waves, V.B.Braginsky, L.P.Grishchuck, A.G.Dooshkevieh, M.B. I.D.Novikov, M.V.Sazhin and Y.B.Zeldovisch – GR.G. Vol.11 No.6 p. 407-408 (1979)

- On the propagation of electromagnetic radiation in the field of a plane gravitational wave, E.Montanari – gr-qc/9806054 (June 11, 1998)

8.8.1   Further reading

- Einstein's Unfinished Symphony by Marcia Bartusiak, ISBN 0-425-18620-2.

- Fundamentals of Interferometric Gravitational Wave Detectors by Peter R. Saulson, ISBN 981-02-1820-6.

- Gravity's Shadow: The Search for Gravitational Waves by Harry Collins, ISBN 0-226-11378-7.

- Traveling at the Speed of Thought by Daniel Kennefick, ISBN 978-0-691-11727-0

8.9 External links

- LIGO Scientific Collaboration web page

- LIGO outreach webpage, with links to summaries of the Collaboration's scientific articles, written for a general public audience

- LIGO Laboratory

- LIGO News blog

- Living LIGO blog: answering questions about LIGO science and being a scientist by LIGO member Amber Stuver

- Advanced LIGO homepage

- Columbia Experimental Gravity

- American Museum of Natural History film and other materials on LIGO

- 40 m Prototype

- Earth-Motion studies A brief discussion of efforts to correct for seismic and human-related activity that contributes to the background signal of the LIGO detectors.

- Caltech's Physics 237-2002 Gravitational Waves by Kip Thorne Video plus notes: Graduate level but does not assume knowledge of General Relativity, Tensor Analysis, or Differential Geometry; Part 1: Theory (10 lectures), Part 2: Detection (9 lectures)

- Caltech Tutorial on Relativity – An extensive description of gravitational waves and their sources.

Chapter 9

Virgo interferometer

"VIRGO" redirects here. For other uses, see Virgo (disambiguation).

The **VIRGO** is a gravitational wave detector in Italy, which started operating in 2007. It is one of a handful of the world's major experiments working towards the observation of gravitational waves. VIRGO is located within the site of EGO (European Gravitational Observatory) at Santo Stefano a Macerata, Cascina, Tuscany.

9.1 Description

VIRGO is a massive Michelson laser interferometer made of two orthogonal arms, each three kilometers long.[2] Due to the multiple reflections between mirrors located at the extremities of each arm, the effective optical length of each arm is extended up to 100 kilometers.

The gravitational wave frequency range sensitivity of VIRGO extends from 10 hertz to 10,000 hertz.[2] This range, as well as the very high sensitivity, should allow detection of gravitational radiation produced by supernovae and the coalescence of binary systems in the Milky Way and in outer galaxies.

In order to reach the extreme sensitivity required, the whole interferometer must remain as optically perfect as possible and is extremely well isolated from seismic disturbances[3] in order to be only sensitive to the gravitational waves. To achieve this, Italian and French project scientists have developed many of the current leading techniques in the fields of high power ultrastable lasers, high reflectivity mirrors, seismic isolation and position and alignment control. To avoid spurious motions of the optical components due to seismic noise; each component is isolated by a 10 m high, very elaborate system of compound pendulums.

Regarding optics, VIRGO uses a new generation of ultrastable lasers and one of the most stable oscillators ever built.[2] A specific optical coating facility was built to produce extremely high quality mirrors combining reflectivity over 99.999%, with nanometer surface control.

Because the presence of any residual gas would disturb the measurements, the light beam must propagate under an ultra high vacuum. Indeed, the two tubes, 3 km long and 1.2 m diameter are the largest ultra high vacuum vessels in Europe and the second largest in the world.

The environment of the Virgo interferometer is quieter than that of a spacecraft orbiting the earth. The signals are detected, recorded and pre-analysed through an on-line computing system. These data will then be made available to the scientific community for further analysis.

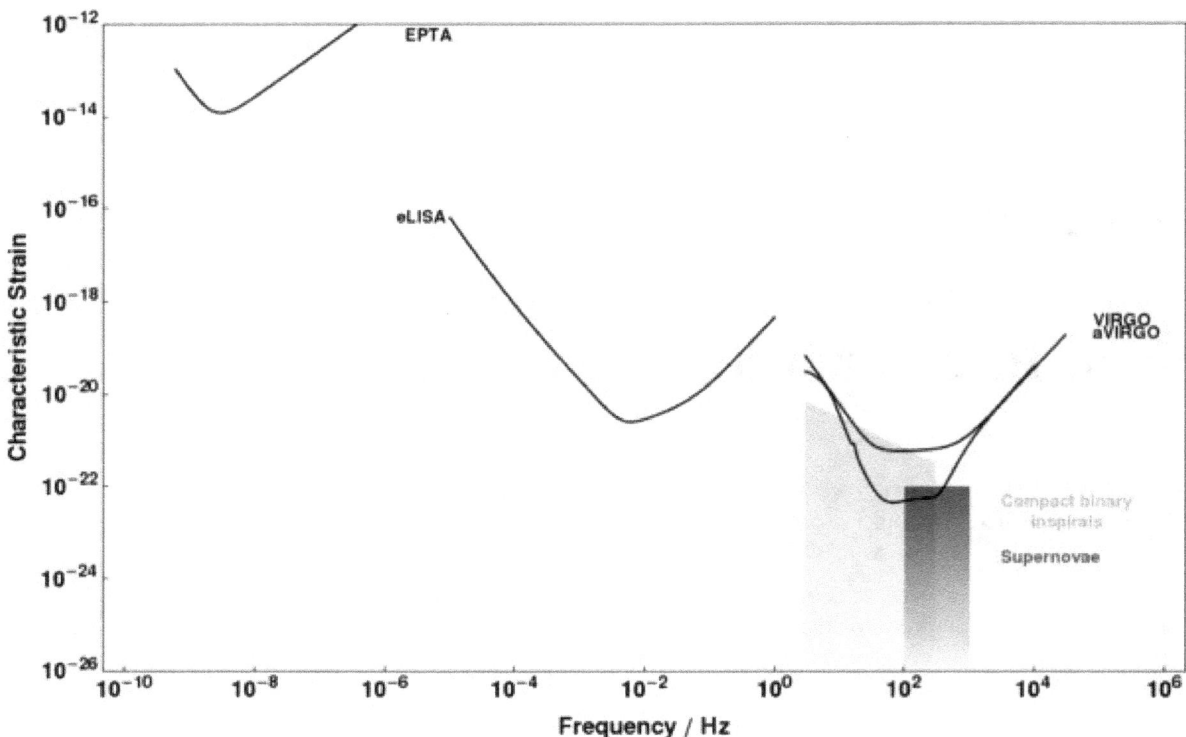

Detector noise curves for Virgo and Advanced Virgo as a function of frequency. They lie above the bands for space-borne detectors like the evolved Laser Interferometer Space Antenna (eLISA) and pulsar timing arrays such as the European Pulsar Timing Array (EPTA). The characteristic strain of potential astrophysical sources are also shown. To be detectable the characteristic strain of a signal must be above the noise curve.[11]

9.2 History

VIRGO, the construction of which was completed in June 2003, started its first science run in May 2007 and it is currently undergoing upgrades to improve sensitivity.

The project has 48 staffers and runs on an annual 10 million euro budget.[4]

VIRGO — view of west tube from the complex gates

9.3 References

[1] Moore, Christopher; Cole, Robert; Berry, Christopher (19 July 2013). "Gravitational Wave Detectors and Sources". Retrieved 20 April 2014.

[2] "The VIRGO interferometer". European Gravitational Observatory. Retrieved April 7, 2013.

[3] http://ieeexplore.ieee.org/xpl/login.jsp?tp=&arnumber=4998833&url=http%3A%2F%2Fieeexplore.ieee.org%2Fxpls%2Fabs_all.jsp%3Farnumber%3D4998833

[4] http://www.icra.it/MG/mg12/talks/gw2_calloni.ppt Commissioning status of the Virgo interferometer, 2009 (Power Point)

9.4   External links

- Description on EGO's website

- Virgo's homepage

- Advanced Virgo Technical Design Report

Coordinates: 43°37′53″N 10°30′18″E / 43.63139°N 10.50500°E

Chapter 10

Evolved Laser Interferometer Space Antenna

The **Evolved Laser Interferometer Space Antenna (eLISA)**, previously called **LISA**, is a proposed European Space Agency mission designed to detect and accurately measure gravitational waves[3] — tiny ripples in the fabric of space-time — from astronomical sources.[4] The mission would evolve from the ESA LISA Pathfinder technology research demonstrator scheduled for launch in November 27, 2015 into a full scale gravitational wave observatory.[5][6] eLISA would be the first dedicated space-based gravitational wave detector. It aims to measure gravitational waves directly by using laser interferometry. The LISA concept has a constellation of three spacecraft, arranged in an equilateral triangle with million kilometre arms (5 million km for classic LISA, 1 million km for eLISA) flying along an Earth-like heliocentric orbit. The distance between the satellites is precisely monitored to detect a passing gravitational wave.[3]

The LISA project (Laser Interferometer Space Antenna) was previously a joint effort between the United States space agency NASA and the European Space Agency ESA. However, on April 8, 2011, NASA announced that it would be unable to continue its LISA partnership with the European Space Agency,[7] due to funding limitations.[8] ESA has therefore revised the mission's concept to fit into a European-only cost envelope. The scaled down design was initially known as the **New Gravitational-wave Observatory (NGO)** for ESA's L1 mission selection.[9] Following this unsuccessful application, the name was changed to eLISA.[10] The project was chosen as the L3 mission within the ESA Cosmic Vision Program, with a tentative launch date in 2034.[2]

A LISA-like mission is designed to directly observe gravitational waves, which are distortions of space-time travelling at the speed of light. Passing gravitational waves alternately squeeze and stretch objects by a tiny amount. Gravitational waves are caused by energetic events in the universe and, unlike any other radiation, can pass unhindered by intervening mass. Launching eLISA will add a new sense to scientists' perception of the universe and enable them to listen to a world that is invisible with light.[11][12]

Potential sources for signals are merging massive black holes at the centre of galaxies,[13] massive black holes [14] orbited by small compact objects, known as extreme mass ratio inspirals, binaries of compact stars in our Galaxy,[15] and possibly other sources of cosmological origin, such as the very early phase of the Big Bang,[16] and speculative astrophysical objects like cosmic strings and domain boundaries.[17]

10.1 Mission description

The LISA/eLISA Mission's primary objective is to detect and measure gravitational waves produced by compact binary systems and mergers of supermassive black holes. LISA/eLISA will observe gravitational waves by measuring differential changes in the length of its arms, as sensed by laser interferometry.[18] Each of the LISA spacecraft contains two telescopes, two lasers and two test masses, arranged in two optical assemblies pointed at the other two spacecraft. This forms Michelson-like interferometers, each centred on one of the spacecraft, with the platinum-gold test masses defining the ends of the arms.[19] The entire arrangement, which is ten times larger than the orbit of the Moon, will be placed in

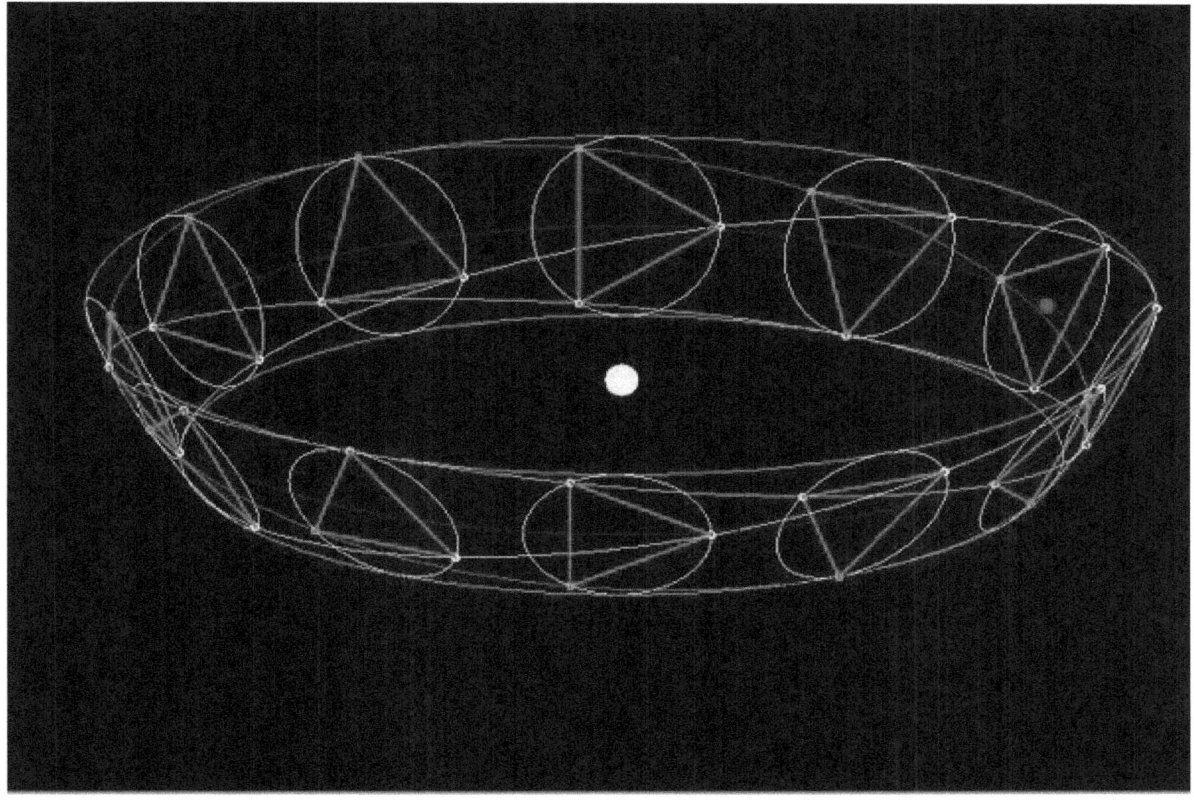

LISA configuration.

solar orbit at the same distance from the Sun as the Earth, but trailing the Earth by 20 degrees, and with the orbital planes of the 3 sciencecraft inclined relative to the ecliptic by about 0.33 degree, which results in the plane of the triangular sciencecraft formation being tilted 60 degrees from the plane of the ecliptic.[18] The mean linear distance between the constellation and the Earth will be 50 million kilometers.[20]

To eliminate non-gravitational forces such as light pressure and solar wind on the test masses, each spacecraft is constructed as a zero-drag satellite, and effectively floats around the masses, using capacitive sensing to determine their position relative to the spacecraft, and very precise thrusters to keep itself centered around them.[21]

10.2   LISA Pathfinder

An ESA test mission called LISA Pathfinder (LPF) will prove LISA/eLISA's key technologies in space. LPF consists of a single spacecraft with one of the LISA/eLISA interferometer arms shortened to about 38 cm, so that it fits inside a single spacecraft. LPF will be launched in 2015.[22] [23]

10.3   Science

Gravitational-wave astronomy seeks to use direct measurements of gravitational waves to study astrophysical systems and to test Einstein's theory of gravity. The existence of gravitational waves has been confirmed from observations of the decreasing orbital periods of several binary pulsars, such as the famous Hulse–Taylor binary pulsar.[25] However, gravitational waves have not yet been directly detected on Earth because of their extremely small effect on matter.

Observing gravitational waves requires two things: a strong source of gravitational waves —— such as the merger of two

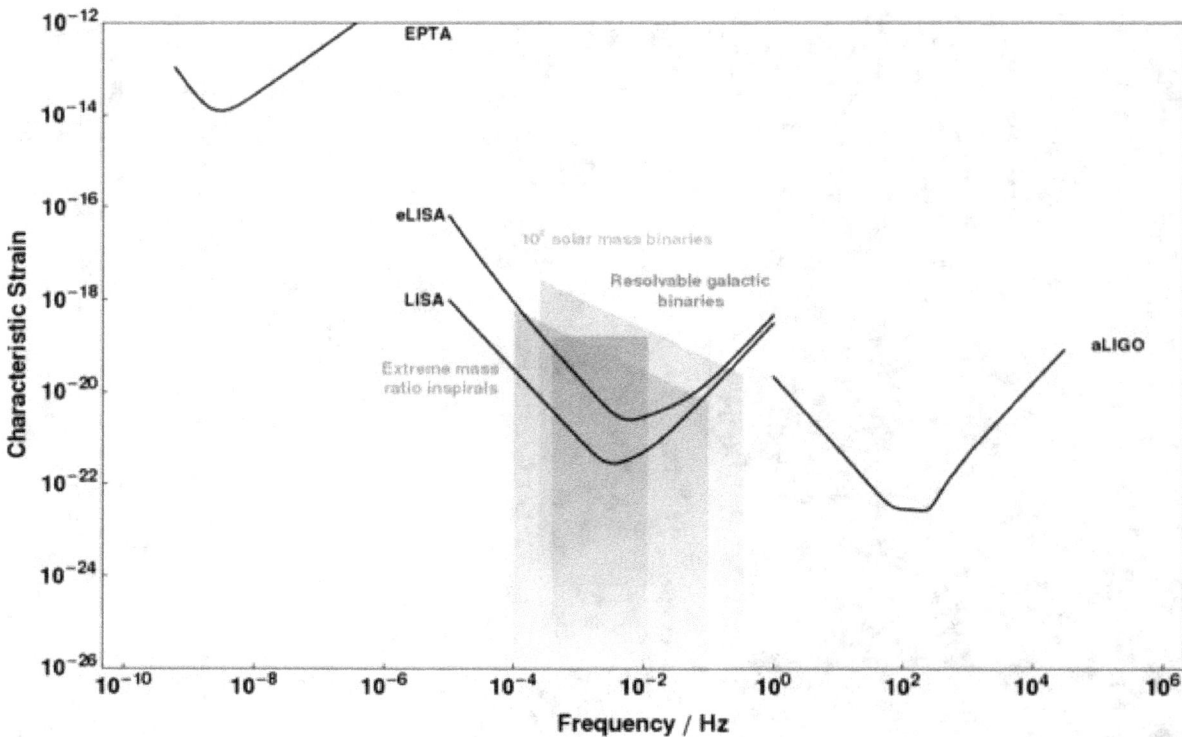

Detector noise curves for LISA and eLISA as a function of frequency. They lie in between the bands for ground-based detectors like advanced LIGO (aLIGO) and pulsar timing arrays such as the European Pulsar Timing Array (EPTA). The characteristic strain of potential astrophysical sources are also shown. To be detectable the characteristic strain of a signal must be above the noise curve.[24]

black holes —— and extremely high detection sensitivity. A LISA-like instrument should be able to measure relative displacements with a resolution of 20 picometers —— less than the diameter of a helium atom —— over a distance of a million kilometres, yielding a strain sensitivity of better than 1 part in 10^{20} in the low-frequency band about a millihertz.

A LISA-like detector is sensitive to the low-frequency band of the gravitational-wave spectrum, which contains many astrophysically interesting sources.[26] Such a detector would observe signals from binary stars within our galaxy (the Milky Way);[27][28] signals from binary supermassive black holes in other galaxies;[29] and extreme-mass-ratio inspirals and bursts produced by a stellar-mass compact object orbiting a supermassive black hole.[30][31] There are also more speculative signals such as signals from cosmic strings and primordial gravitational waves generated during cosmological inflation.[32]

10.4   Other gravitational-wave experiments

Previous searches for gravitational waves in space were conducted for short periods by planetary missions that had other primary science objectives (such as Cassini–Huygens), using microwave Doppler tracking to monitor fluctuations in the Earth-spacecraft distance. By contrast, LISA is a dedicated mission that will use laser interferometry to achieve a much higher sensitivity. Other gravitational wave antennas, such as LIGO, VIRGO, and GEO 600, are already in operation on Earth, but their sensitivity at low frequencies is limited by the largest practical arm lengths, by seismic noise, and by interference from nearby moving masses. Thus, LISA and ground detectors are complementary rather than competitive, much like astronomical observatories in different electromagnetic bands (e.g., ultraviolet and infrared).

10.5   History

The first design studies for gravitational wave detector to be flown in space were performed in the 1980s under the name LAGOS (Laser Antena for Gravitational radiation Observation in Space). LISA was first proposed as a mission to ESA in the early 1990s. First as a candidate for the M3-cycle, and later as 'cornerstone mission' for the 'Horizon 2000 plus' program. As the decade progressed, the design was refined to a triangular configuration of three spacecraft with three 5 million kilometer arms. This mission was pitched as a joint mission between ESA and NASA in 1997.[33]

In the 2000s the joint ESA/NASA LISA mission was identified as a candidate for the 'L1' slot in ESA's Cosmic Vision 2015-2025 programme. However, due to budget cuts, NASA announced in early 2011 that it would not be contributing to any of ESA's L-class missions. ESA nonetheless decided to push the program forward, and instructed the L1 candidate missions to present reduced cost versions that could be flown within ESA's budget. A reduced version of LISA was designed with only two 1 million kilometer arms under the name NGO (New/Next Gravitational wave Observatory). Despite NGO being ranked highest in terms of scientific potential, ESA decided to fly Jupiter Icy Moon Explorer (JUICE) as its L1 mission. One of the main concerns was, that the LISA Pathfinder mission had been experiencing technical delays, making it uncertain if the technology would be ready for the projected L1 launch date.[33]

Soon afterwards, ESA announced it would be selecting themes for its L2 and L3 mission slots. A theme called "the Gravitational Universe" was formulated with the reduced NGO rechristened eLISA as a straw-man mission.[34] In November 2013, ESA announced that it selected "the Gravitational Universe" for its L3 mission slot (expected launch in 2032).[35]

10.6   See also

- Cosmic Vision program - ESA

- Beyond Einstein program - NASA

- Big Bang Observer - proposed LISA successor

- DECIGO - proposed Japanese equivalent

10.7   External links

- "eLISA Consortium portal". Retrieved 2013-11-13.

- "ESA LISA homepage". Retrieved 2010-08-13.

- "LISA Pathfinder mission". Retrieved 2013-05-22.

- "LISA International Science Team website". Retrieved 2010-08-13.

- "NASA LISA homepage". Retrieved 2010-08-13.

10.8   References

[1] "Selected: The gravitational universe; ESA decides on next large mission concepts". eLISA Consortium. Retrieved 29 November 2013.

[2] "ESA's new vision to study the invisible universe". ESA. Retrieved 29 November 2013.

[3] "eLISA, The First Gravitational Wave Observatory in Space". eLISA Consortium. Retrieved 12 November 2013.

[4] "eLISA, Partners and Contacts". eLISA Consortium. Retrieved 12 November 2013.

[5] "ESA Lisa Pathfinder overview". ESA. 6 June 2013. Retrieved 12 November 2013.

[6] "eLISA: LPF". eLISA Consortium. Retrieved 12 November 2013.

[7] "LISA on the NASA website". NASA. Retrieved 12 November 2013.

[8] "President's FY12 Budget Request". NASA/US Federal Government. Retrieved 4 Mar 2011.

[9] Amaro-Seoane, Pau; Aoudia, Sofiane; Babak, Stanislav; Binétruy, Pierre; Berti, Emanuele; Bohé, Alejandro; Caprini, Chiara; Colpi, Monica; Cornish, Neil J; Danzmann, Karsten; Dufaux, Jean-François; Gair, Jonathan; Jennrich, Oliver; Jetzer, Philippe; Klein, Antoine; Lang, Ryan N; Lobo, Alberto; Littenberg, Tyson; McWilliams, Sean T; Nelemans, Gijs; Petiteau, Antoine; Porter, Edward K; Schutz, Bernard F; Sesana, Alberto; Stebbins, Robin; Sumner, Tim; Vallisneri, Michele; Vitale, Stefano; Volonteri, Marta; Ward, Henry (21 June 2012). "Low-frequency gravitational-wave science with eLISA/NGO". *Classical and Quantum Gravity* **29** (12): 124016. arXiv:1202.0839. Bibcode:2012CQGra..29l4016A. doi:10.1088/0264-9381/29/12/124016.

[10] Selected: The Gravitational Universe ESA decides on next Large Mission Concepts.

[11] "eLISA: Science Context 2028". eLISA Consortium. Retrieved 15 November 2013.

[12] "Gravitational-Wave Detetectors Get Ready to Hunt for the Big Bang". Scientific American. 17 September 2013.

[13] "eLISA whitepaper, sec. 5.2 p 40, arXiV:1201.3621v1". eLISA consortium. 17 Jan 2012.

[14] "eLISA whitepaper, sec. 4.3, p. 25, arXiV:1201.3621v1". eLISA consortium. 17 Jan 2012.

[15] "eLISA whitepaper, sec 3.3, p 11, arXiV:1201.3621v1". eLISA consortium. 17 Jan 2012.

[16] "eLISA whitepaper, sec 7.2, p 59, arXiV:1201.3621v1". eLISA consortium. 17 Jan 2012.

[17] "eLISA whitepaper, sec 1.1, p 5, arXiV:1201.3621v1". eLISA consortium. 17 Jan 2012.

[18] "eLISA: the mission concept". eLISA Consortium. Retrieved 12 November 2013.

[19] "eLISA: distance measurement". eLISA Consortium. Retrieved 12 November 2013.

[20] "eLISA: key features". eLISA Consortium. Retrieved 12 November 2013.

[21] "eLISA: dragfree operation". eLISA Consortium. Retrieved 12 November 2013.

[22] "eLISA: Lisa Pathfinder". eLISA Consortium. Retrieved 12 November 2013.

[23] "ESA: Lisa Pathfinder overview". European Space Agency. Retrieved 12 November 2013.

[24] Moore, Christopher; Cole, Robert; Berry, Christopher (19 July 2013). "Gravitational Wave Detectors and Sources". Retrieved 14 April 2014.

[25] Stairs, Ingrid H. "Testing General Relativity with Pulsar Timing". *Living Reviews in Relativity* **6**. arXiv:astro-ph/0307536. Bibcode:2003LRR.....6....5S. doi:10.12942/lrr-2003-5.

[26] Amaro-Seoane, Pau; Aoudia, Sofiane; Babak, Stanislav; Binétruy, Pierre; Berti, Emanuele; Bohé, Alejandro; Caprini, Chiara; Colpi, Monica; Cornish, Neil J; Danzmann, Karsten; Dufaux, Jean-François; Gair, Jonathan; Jennrich, Oliver; Jetzer, Philippe; Klein, Antoine; Lang, Ryan N; Lobo, Alberto; Littenberg, Tyson; McWilliams, Sean T; Nelemans, Gijs; Petiteau, Antoine; Porter, Edward K; Schutz, Bernard F; Sesana, Alberto; Stebbins, Robin; Sumner, Tim; Vallisneri, Michele; Vitale, Stefano; Volonteri, Marta; Ward, Henry (21 June 2012). "Low-frequency gravitational-wave science with eLISA/NGO". *Classical and Quantum Gravity* **29** (12): 124016. arXiv:1202.0839. Bibcode:2012CQGra..29l4016A. doi:10.1088/0264-9381/29/12/124016.

[27] Nelemans, Gijs (7 May 2009). "The Galactic gravitational wave foreground". *Classical and Quantum Gravity* **26** (9): 094030. arXiv:0901.1778. Bibcode:2009CQGra..26i4030N. doi:10.1088/0264-9381/26/9/094030.

[28] Stroeer, A; Vecchio, A (7 October 2006). "The LISA verification binaries". *Classical and Quantum Gravity* **23** (19): S809–S817. arXiv:astro-ph/0605227. Bibcode:2006CQGra..23S.809S. doi:10.1088/0264-9381/23/19/S19.

[29] Flanagan, Éanna É. "Measuring gravitational waves from binary black hole coalescences. I. Signal to noise for inspiral, merger, and ringdown".*Physical Review D*57(8): 4535–4565.arXiv:gr-qc/9701039.Bibcode:1998PhRvD..57.4535F.doi:10.1103/

[30] Amaro-Seoane, Pau; Gair, Jonathan R; Freitag, Marc; Miller, M Coleman; Mandel, Ilya; Cutler, Curt J; Babak, Stanislav (7 September 2007). "Intermediate and extreme mass-ratio inspirals—astrophysics, science applications and detection using LISA". *Classical and Quantum Gravity* **24** (17): R113–R169. arXiv:astro-ph/0703495. Bibcode:2007CQGra..24R.113A. doi:10.1088/0264-9381/24/17/R01.

[31] Berry, C. P. L.; Gair, J. R. (12 September 2013). "Expectations for extreme-mass-ratio bursts from the Galactic Centre". *Monthly Notices of the Royal Astronomical Society* **435** (4): 3521–3540. arXiv:1307.7276. Bibcode:2013MNRAS.435.3521B. doi:10.1093/mnras/stt1543.

[32] Binétruy, Pierre; Bohé, Alejandro; Caprini, Chiara; Dufaux, Jean-François (13 June 2012). "Cosmological backgrounds of gravitational waves and eLISA/NGO: phase transitions, cosmic strings and other sources". *Journal of Cosmology and Astroparticle Physics* **2012** (06): 027–027. arXiv:1201.0983. Bibcode:2012JCAP...06..027B. doi:10.1088/1475-7516/2012/06/027.

[33]   and

[34] Danzmann, Karsten; The eLISA Consortium (2013). "The Gravitational Universe" (PDF). Retrieved 15 April 2014.

[35] "Selected: The Gravitational Universe ESA decides on next Large Mission Concepts". Max Planck Institute for Gravitational Physics.

Chapter 11

Linearized gravity

Linearized gravity is an approximation scheme in general relativity in which the nonlinear contributions from the spacetime metric are ignored, simplifying the study of many problems while still producing useful approximate results.

11.1　The method

In linearized gravity the metric tensor, g , of spacetime is treated as a sum of an exact solution of Einstein's equations (often Minkowski spacetime) and a perturbation h .

$$g = \eta + h$$

where η is the nondynamical background metric that is being perturbed about, and h represents the deviation of the true metric (g) from flat spacetime.

The perturbation is treated using the methods of perturbation theory, "linearized" by ignoring all terms of order higher than one (quadratic in h , cubic in h etc...) in the perturbation.

11.2　Applications

The Einstein field equations (EFE), being nonlinear in the metric, are difficult to solve exactly and the above perturbation scheme allows linearised Einstein field equations to be obtained. These equations are linear in the metric, and the sum of two solutions of the linearized EFE is also a solution. The idea of 'ignoring the nonlinear part' is thus encapsulated in this linearization procedure.

The method is used to derive the Newtonian limit, including the first corrections, much like for a derivation of the existence of gravitational waves that led, after quantization, to gravitons. This is why the conceptual approach of linearized gravity is the canonical one in particle physics, string theory, and more generally quantum field theory where classical (bosonic) fields are expressed as coherent states of particles.

This approximation is also known as the weak-field approximation as it is only valid for h very small.

11.2.1　Weak-field approximation

In a weak-field approximation, the gauge symmetry is associated with diffeomorphisms with small "displacements" (diffeomorphisms with large displacements obviously violate the weak field approximation), which has the exact form (for infinitesimal transformations)

$$\delta_{\vec{\xi}} h = \delta_{\vec{\xi}} g - \delta_{\vec{\xi}} \eta = \mathcal{L}_{\vec{\xi}} g = \mathcal{L}_{\vec{\xi}} \eta + \mathcal{L}_{\vec{\xi}} h = \left[\xi_{\nu;\mu} + \xi_{\mu;\nu} + \xi^{\alpha} h_{\mu\nu;\alpha} + \xi^{\alpha}_{;\mu} h_{\alpha\nu} + \xi^{\alpha}_{;\nu} h_{\mu\alpha} \right] dx^{\mu} \otimes dx^{\nu}$$

Where $\mathcal{L}$ is the Lie derivative and we used the fact that η does not transform (by definition). Note that we are raising and lowering the indices with respect to η and not g and taking the covariant derivatives (Levi-Civita connection) with respect to η. This is the standard practice in linearized gravity. The way of thinking in linearized gravity is this: the background metric η is the metric and h is a field propagating over the spacetime with this metric.

In the weak field limit, this gauge transformation simplifies to

$$\delta_{\vec{\xi}} h_{\mu\nu} \approx \left(\mathcal{L}_{\vec{\xi}} \eta \right)_{\mu\nu} = \xi_{\nu;\mu} + \xi_{\mu;\nu}$$

The weak-field approximation is useful in finding the values of certain constants, for example in the Einstein field equations and in the Schwarzschild metric.

11.3   Linearised Einstein field equations

The **linearised Einstein field equations** (linearised EFE) are an approximation to Einstein's field equations that is valid for a weak gravitational field and is used to simplify many problems in general relativity and to discuss the phenomena of gravitational radiation. The approximation can also be used to derive Newtonian gravity as the weak-field approximation of Einsteinian gravity.

The equations are obtained by assuming the spacetime metric is only slightly different from some baseline metric (usually a Minkowski metric). Then the difference in the metrics can be considered as a field on the baseline metric, whose behaviour is approximated by a set of linear equations.

11.3.1   Derivation for the Minkowski metric

Starting with the metric for a spacetime in the form

$$g_{ab} = \eta_{ab} + h_{ab}$$

where η_{ab} is the Minkowski metric and h_{ab} — sometimes written as $\epsilon\,\gamma_{ab}$ — is the deviation of g_{ab} from it. h must be negligible compared to $\eta : |h_{\mu\nu}| \ll 1$ (and similarly for all derivatives of h). Then one ignores all products of h (or its derivatives) with h or its derivatives (equivalent to ignoring all terms of higher order than 1 in ϵ). It is further assumed in this approximation scheme that all indices of h and its derivatives are raised and lowered with η .

The metric h is clearly symmetric, since g and η are. The consistency condition $g_{ab} g^{bc} = \delta_a{}^c$ shows that

$$g^{ab} = \eta^{ab} - h^{ab}$$

The Christoffel symbols can be calculated as

$$2\Gamma^{a}_{bc} = (h^{a}{}_{b,c} + h^{a}{}_{c,b} - h_{bc,}{}^{a})$$

where $h_{bc,}{}^{a} \overset{\text{def}}{=} \eta^{ar} h_{bc,r}$, and this is used to calculate the Riemann tensor:

$$2R^{a}{}_{bcd} = 2(\Gamma^{a}_{bd,c} - \Gamma^{a}_{bc,d}) = \eta^{ae}(h_{eb,dc} + h_{ed,bc} - h_{bd,ec} - h_{eb,cd} - h_{ec,bd} + h_{bc,ed}) =$$

$$= \eta^{ae}(h_{ed,bc} - h_{bd,ec} - h_{ec,bd} + h_{bc,ed}) = h^a_{d,bc} - h_{bd,}{}^a{}_c + h_{bc,}{}^a{}_d - h^a{}_{c,bd}$$

Using $R_{bd} = \delta^c{}_a R^a{}_{bcd}$ gives

$$2R_{bd} = h^r_{d,br} + h^r_{b,dr} - h_{,bd} - h_{bd,rs}\eta^{rs}$$

For Ricci scalar we have:

$$R = R_{bd}\eta^{bd} = h^{ab}_{,ab} - \Box h$$

Then the linearized Einstein equations are

$$8\pi T_{bd} = R_{bd} - R_{ac}\eta^{ac}\eta_{bd}/2$$

or

$$8\pi T_{bd} = (h^r_{d,br} + h^r_{b,dr} - h_{,bd} - h_{bd,r}{}^r - h^r_{s,r}{}^s\eta_{bd})/2 + (h_{,a}{}^a\eta_{bd} + h_{ac,r}{}^r\eta^{ac}\eta_{bd})/4$$

Or, equivalently:

$$8\pi(T_{bd} - T_{ac}\eta^{ac}\eta_{bd}/2) = R_{bd}$$
$$16\pi(T_{bd} - T_{ac}\eta^{ac}\eta_{bd}/2) = h^r_{d,br} + h^r_{b,dr} - h_{,bd} - h_{bd,rs}\eta^{rs}$$

11.4   With a coordinate condition

If one uses the Lorentz invariant harmonic coordinate condition

$$h_{\alpha\beta,\gamma}\eta^{\beta\gamma} = \frac{1}{2}h_{\beta\gamma,\alpha}\eta^{\beta\gamma},$$

then the last form above of the linearized Einstein equation simplifies to

$$16\pi(T_{bd} - T_{ac}\eta^{ac}\eta_{bd}/2) = -h_{bd,rs}\eta^{rs}.$$

To solve it, this can be rewritten as

$$\Delta h_{bd} = \frac{-16\pi G}{c^4}(T_{bd} - T_{ac}\eta^{ac}\eta_{bd}/2) + \frac{\partial^2 h_{bd}}{c^2\partial t^2}$$

where Δ is the Laplacian on a spatial slice. If the stress-energy changes slowly (velocities are low compared to c), then this gives

$$h_{bd}(r) = \frac{-1}{4\pi}\int\left(\frac{-16\pi G}{c^4}(T_{bd}(s) - T_{ac}(s)\eta^{ac}\eta_{bd}/2) + \frac{\partial^2 h_{bd}(s)}{c^2\partial t^2}\right)\frac{1}{|r-s|}d^3s$$

as a generalization of the Newtonian formula for gravitational potential. This is solved iteratively by first replacing the second time derivative by zero and then inserting the h so obtained repeatedly until convergence.

11.5   Applications

The linearised EFE are used primarily in the theory of gravitational radiation, where the gravitational field far from the source is approximated by these equations.

11.6   See also

- Correspondence principle

- Gravitoelectromagnetism

- Lanczos tensor

- Parameterized post-Newtonian formalism

- Quasinormal mode

11.7   References

- Stephani, Hans (1990). *General Relativity: An Introduction to the Theory of the Gravitational Field,*. Cambridge: Cambridge University Press. ISBN 0-521-37941-5.

- Adler, Ronald; Bazin, Maurice' & Schiffer, Menahem (1965). *Introduction to General Relativity*. New York: McGraw-Hill. ISBN 0-07-000423-4.

Chapter 12

Quadrupole formula

In general relativity, the **quadrupole formula** describes rate at which gravitational waves are emitted from a system of masses based on the change of the (mass) quadrupole moment. The formula reads

$$\bar{h}_{ij}(t,r) = \frac{2G}{c^4 r} \ddot{I}_{ij}(t-r),$$

where $\bar{h}_{ij}$ is the (spatial part of) the trace reversed perturbation of the metric (i.e. the gravitational wave), and I_{ij} is the mass quadrupole moment.[1]

The formula was first obtained by Albert Einstein in 1916. After a long history of debate on its physical correctness, observations of energy loss due to gravitational radiation in the Hulse–Taylor binary discovered in 1974 confirmed the result, with agreement up to 0.2 percent (by 2005).[2]

12.1   References

[1] Carroll, Sean M. *Spacetime and Geometry*. Pearson/Addison Wesley. pp. 300–307. ISBN 0805387323.

[2] Poisson, Eric; Will, Clifford M. *Gravity:Newtonian, Post-Newtonian, Relativistic*. Cambridge University Press. pp. 550–563. ISBN 9781107032866.

Chapter 13

SQUID

For other uses, see Squid (disambiguation).

A **SQUID** (for **superconducting quantum interference device**) is a very sensitive magnetometer used to measure

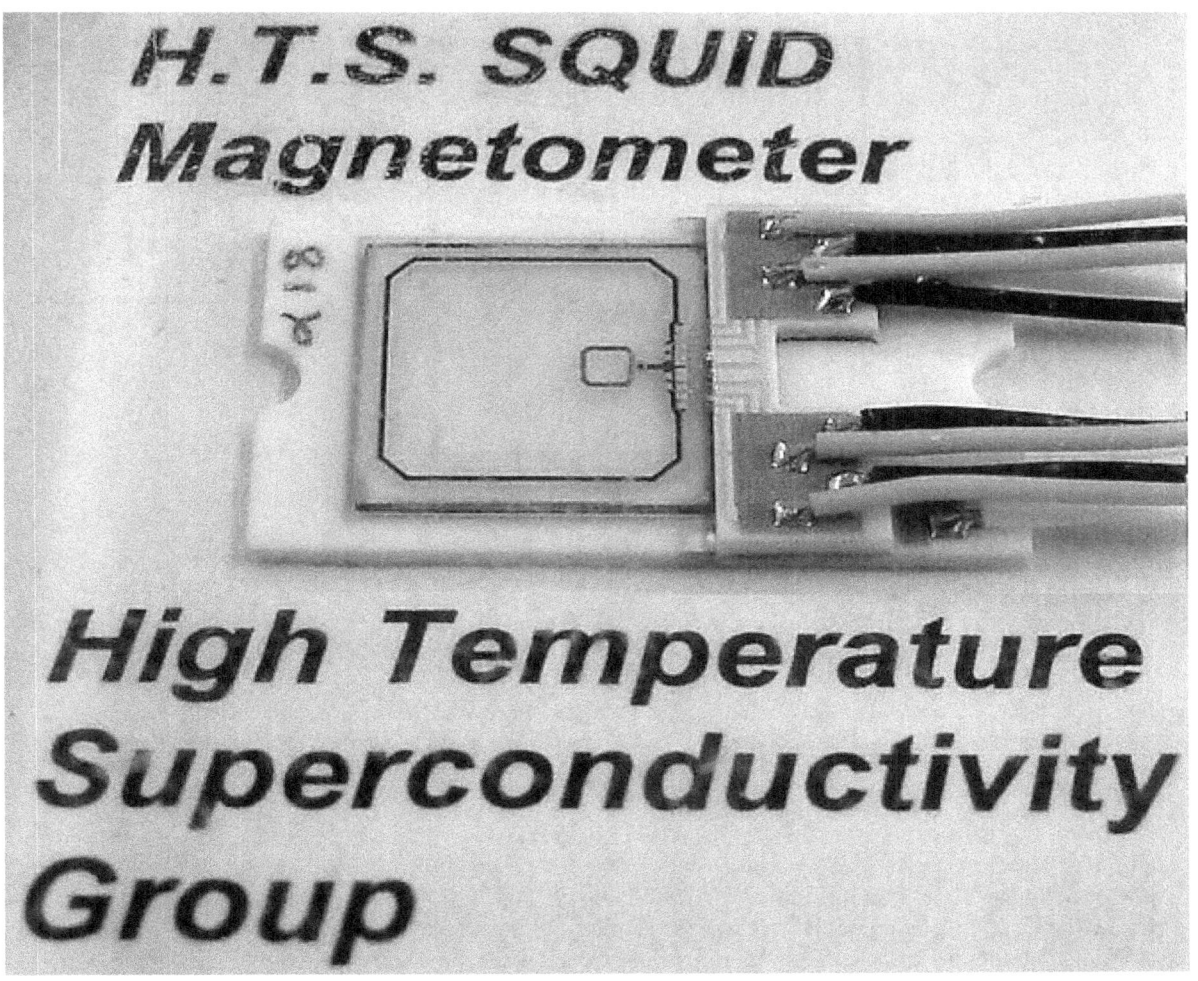

Sensing element of the SQUID

extremely subtle magnetic fields, based on superconducting loops containing Josephson junctions.

SQUIDs are sensitive enough to measure fields as low as 5 aT (5×10^{-18} T) within a few days of averaged measurements.[1] Their noise levels are as low as 3 fT·Hz$^{-\frac{1}{2}}$.[2] For comparison, a typical refrigerator magnet produces 0.01 tesla (10^{-2} T), and some processes in animals produce very small magnetic fields between 10^{-9} T and 10^{-6} T. Recently invented SERF atomic magnetometers are potentially more sensitive and do not require cryogenic refrigeration but are orders of magnitude larger in size (~1 cm^3) and must be operated in a near-zero magnetic field.

13.1 History and design

There are two main types of SQUID: direct current (DC) and radio frequency (RF). RF SQUIDs can work with only one Josephson junction (superconducting tunnel junction), which might make them cheaper to produce, but are less sensitive.

13.1.1 DC SQUID

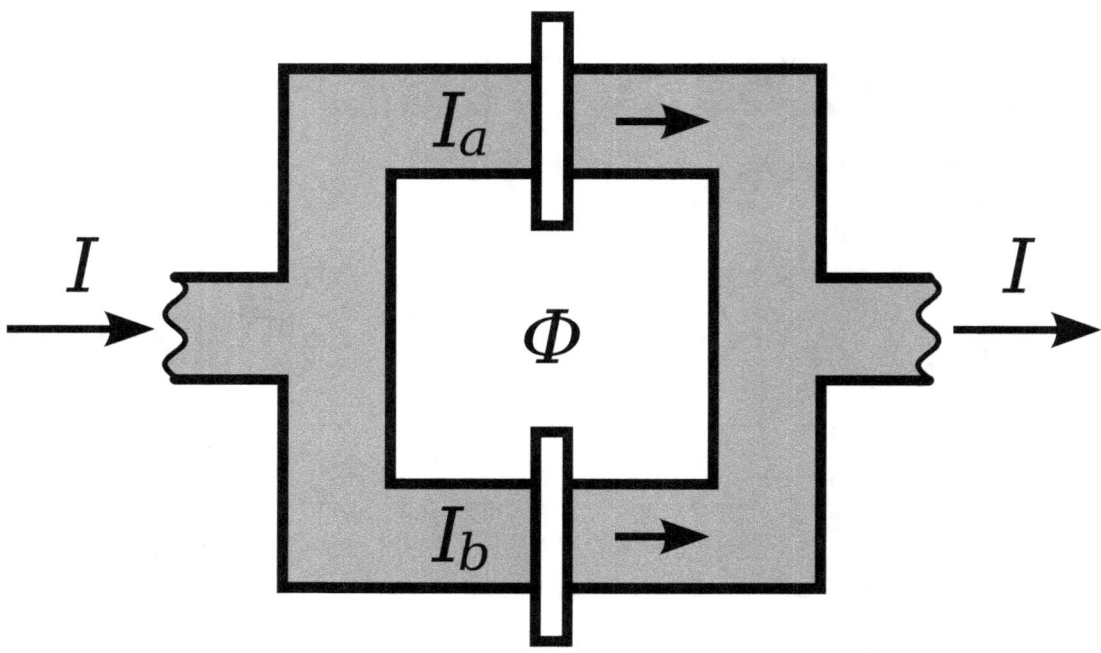

Diagram of a DC SQUID. The current I enters and splits into the two paths, each with currents I_a and I_b. The thin barriers on each path are Josephson junctions, which together separate the two superconducting regions. Φ represents the magnetic flux threading the DC SQUID loop.

The DC SQUID was invented in 1964 by Robert Jaklevic, John J. Lambe, James Mercereau, and Arnold Silver of Ford Research Labs[3] after Brian David Josephson postulated the Josephson effect in 1962, and the first Josephson junction was made by John Rowell and Philip Anderson at Bell Labs in 1963.[4] It has two Josephson junctions in parallel in a superconducting loop. It is based on the DC Josephson effect. In the absence of any external magnetic field, the input current I splits into the two branches equally. If a small external magnetic field is applied to the superconducting loop, a screening current, I_s, begins circulating in the loop that generates a magnetic field canceling the applied external flux. The induced current is in the same direction as I in one of the branches of the superconducting loop, and is opposite to I in the other branch; the total current becomes $I/2 + I_s$ in one branch and $I/2 - I_s$ in the other. As soon as the current in either branch exceeds the critical current, I_c, of the Josephson junction, a voltage appears across the junction.

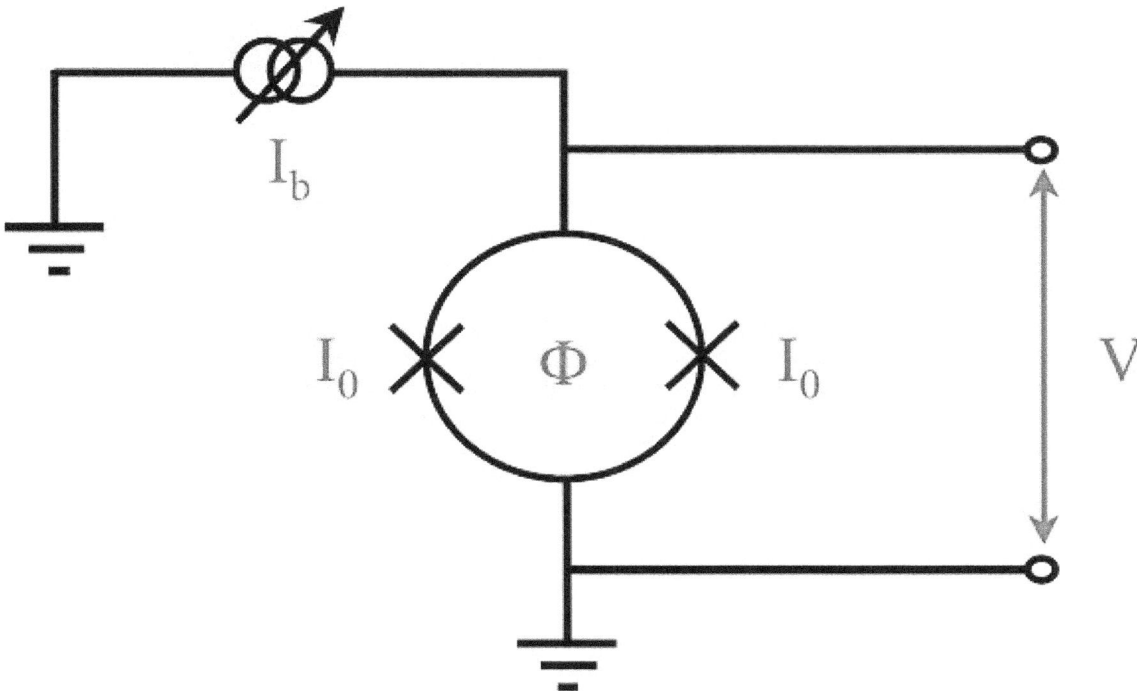

Electrical schematic of a SQUID where Ib is the bias current, I_0 is the critical current of the SQUID, Φ is the flux threading the SQUID and V is the voltage response to that flux. The X-symbols represent Josephson junctions.

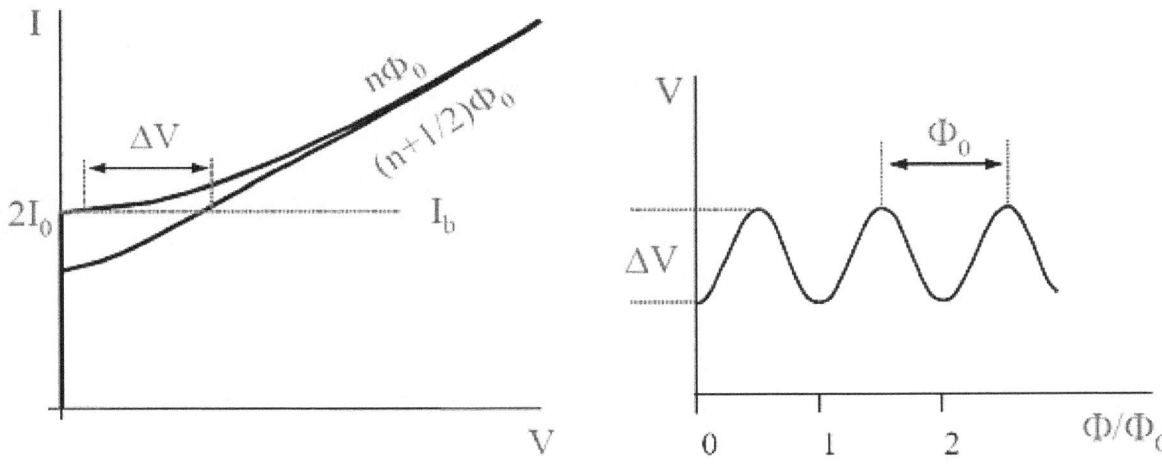

Left: Plot of current vs. voltage for a SQUID. Upper and lower curves correspond to $n\Phi_0$ and $(n+1/2)\Phi_0$ respectively. Right: Periodic voltage response due to flux through a SQUID. The periodicity is equal to one flux quantum, Φ_0

Now suppose the external flux is further increased until it exceeds $\Phi_0/2$, half the magnetic flux quantum. Since the flux enclosed by the superconducting loop must be an integer number of flux quanta, instead of screening the flux the SQUID now energetically prefers to increase it to Φ_0 . The screening current now flows in the opposite direction. Thus the screening current changes direction every time the flux increases by half integer multiples of Φ_0 . Thus the critical current oscillates as a function of the applied flux. If the input current is more than I_c , then the SQUID always operates in the resistive mode. The voltage in this case is thus a function of the applied magnetic field and the period equal to Φ_0 . Since the current-voltage characteristics of the DC SQUID is hysteretic, a shunt resistance, R is connected across the junction to eliminate the hysteresis (in the case of copper oxide based high-temperature superconductors the junction's own intrinsic resistance is usually sufficient). The screening current is the applied flux divided by the self-inductance of the ring. Thus $\Delta\Phi$ can be estimated as the function of ΔV (flux to voltage converter)[5][6] as follows:

$$\Delta V = R\,\Delta I$$

$$2I = 2\,\Delta\Phi/L, \text{ where } L \text{ is the self inductance of the superconducting ring}$$

$$\Delta V = (R/L)\,\Delta\Phi$$

The discussion in this Section assumed perfect flux quantization in the loop. However, this is only true for big loops with a large self-inductance. According to the relations, given above, this implies also small current and voltage variations. In practice the self-inductance L of the loop is not so large. The general case can be evaluated by introducing a parameter

$$\lambda = \frac{i_c L}{\Phi_0}$$

with i_c the critical current of the SQUID. Usually λ is of order one.[7]

13.1.2 RF SQUID

The RF SQUID was invented in 1965 by Robert Jaklevic, John J. Lambe, Arnold Silver, and James Edward Zimmerman at Ford.[6] It is based on the AC Josephson effect and uses only one Josephson junction. It is less sensitive compared to DC SQUID but is cheaper and easier to manufacture in smaller quantities. Most fundamental measurements in biomagnetism, even of extremely small signals, have been made using RF SQUIDS.[8][9] The RF SQUID is inductively coupled to a resonant tank circuit. Depending on the external magnetic field, as the SQUID operates in the resistive mode, the effective inductance of the tank circuit changes, thus changing the resonant frequency of the tank circuit. These frequency measurements can be easily taken, and thus the losses which appear as the voltage across the load resistor in the circuit are a periodic function of the applied magnetic flux with a period of Φ_0. For a precise mathematical description refer to the original paper by Erné et al.[5][10]

13.1.3 Materials used

The traditional superconducting materials for SQUIDs are pure niobium or a lead alloy with 10% gold or indium, as pure lead is unstable when its temperature is repeatedly changed. To maintain superconductivity, the entire device needs to operate within a few degrees of absolute zero, cooled with liquid helium.

In 2006, proof of concept has be shown for CNT-SQUID sensors build with Aluminum (for the loop) and single walled carbon nanotube (CNT).[11] The sensors is few 100 nm size and operates at 1K or below . Such sensors allows to count spins.[12]

High-temperature SQUID sensors are more recent; they are made of high-temperature superconductors, particularly YBCO, and are cooled by liquid nitrogen which is cheaper and more easily handled than liquid helium. They are less sensitive than conventional *low temperature* SQUIDs but good enough for many applications.

13.2 Uses

The extreme sensitivity of SQUIDs makes them ideal for studies in biology. Magnetoencephalography (MEG), for example, uses measurements from an array of SQUIDs to make inferences about neural activity inside brains. Because SQUIDs can operate at acquisition rates much higher than the highest temporal frequency of interest in the signals emitted by the brain (kHz), MEG achieves good temporal resolution. Another area where SQUIDs are used is magnetogastrography,

A prototype SQUID

which is concerned with recording the weak magnetic fields of the stomach. A novel application of SQUIDs is the magnetic marker monitoring method, which is used to trace the path of orally applied drugs. In the clinical environment SQUIDs are used in cardiology for magnetic field imaging (MFI), which detects the magnetic field of the heart for diagnosis and risk stratification.

Probably the most common commercial use of SQUIDs is in magnetic property measurement systems (MPMS). These are turn-key systems, made by several manufacturers, that measure the magnetic properties of a material sample. This is typically done over a temperature range from that of 300 mK to roughly 400 K.[13] With the decreasing size of SQUID sensors since the last decade, such sensor can equip the tip of an AFM probe. Such device allows simultaneous measurement of roughness of the surface of a sample and the local magnetic flux.[14]

For example, SQUIDs are being used as detectors to perform magnetic resonance imaging (MRI). While high-field MRI uses precession fields of one to several teslas, SQUID-detected MRI uses measurement fields that lie in the microtesla range. In a conventional MRI system, the signal scales as the square of the measurement frequency (and hence precession field): one power of frequency comes from the thermal polarization of the spins at ambient temperature, while the second power of field comes from the fact that the induced voltage in the pickup coil is proportional to the frequency of the precessing magnetization. In the case of untuned SQUID detection of prepolarized spins, however, the NMR signal strength is independent of precession field, allowing MRI signal detection in extremely weak fields, of order the Earth's field. SQUID-detected MRI has advantages over high-field MRI systems, such as the low cost required to build such a system, and its compactness. The principle has been demonstrated by imaging human extremities, and its future application may include tumor screening.[15]

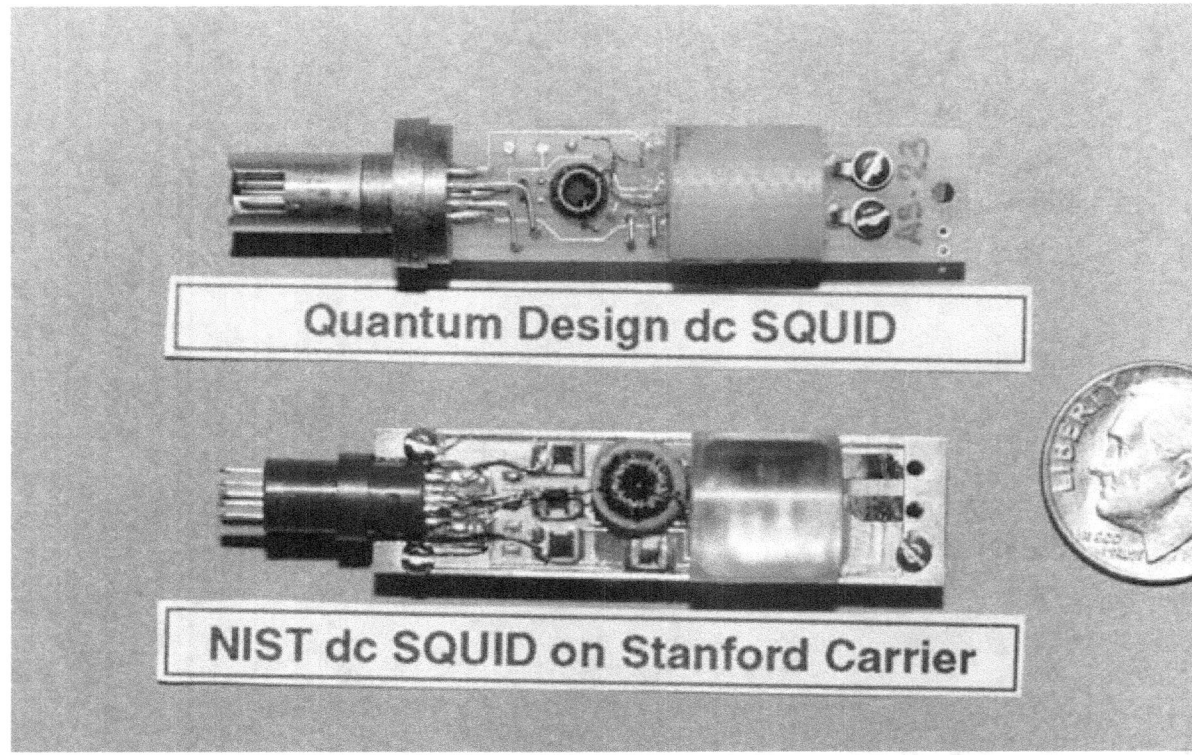

The inner workings of an early SQUID

Another application is the scanning SQUID microscope, which uses a SQUID immersed in liquid helium as the probe. The use of SQUIDs in oil prospecting, mineral exploration, earthquake prediction and geothermal energy surveying is becoming more widespread as superconductor technology develops; they are also used as precision movement sensors in a variety of scientific applications, such as the detection of gravitational waves.[16] A SQUID is the sensor in each of the four gyroscopes employed on Gravity Probe B in order to test the limits of the theory of general relativity.[1]

A modified RF SQUID was used to observe the dynamical Casimir effect for the first time.[17][18]

13.2.1 Proposed uses

It has also been suggested that they might be implemented in a quantum computer.[19]

A potential military application exists for use in anti-submarine warfare as a magnetic anomaly detector (MAD) fitted to maritime patrol aircraft.[20]

13.3 See also

- Macroscopic quantum phenomena

- Geophysics

- Electromagnetism

- Aharonov–Bohm effect

13.4   Notes

[1] Ran, Shannon K'doah (2004). *Gravity Probe B: Exploring Einstein's Universe with Gyroscopes* (PDF). NASA. p. 26.

[2] D. Drung, C. Assmann, J. Beyer, A. Kirste, M. Peters, F. Ruede, and Th. Schurig (2007). "Highly sensitive and easy-to-use SQUID sensors" (PDF). *IEEE Transactions on Applied Superconductivity* **17** (2): 699. Bibcode:2007ITAS...17..699D. doi:10.1109/TASC.2007.897403.

[3] R. C. Jaklevic, J. Lambe, A. H. Silver, and J. E. Mercereau (1964). "Quantum Interference Effects in Josephson Tunneling". *Phys. Rev. Letters* **12** (7): 159–160. Bibcode:1964PhRvL..12..159J. doi:10.1103/PhysRevLett.12.159.

[4] Anderson, P.; Rowell, J. (1963). "Probable Observation of the Josephson Superconducting Tunneling Effect". *Physical Review Letters* **10** (6): 230. Bibcode:1963PhRvL..10..230A. doi:10.1103/PhysRevLett.10.230.

[5] E. du Trémolet de Lacheisserie, D. Gignoux, and M. Schlenker (editors) (2005). *Magnetism: Materials and Applications* **2**. Springer.

[6] J. Clarke and A. I. Braginski (Eds.) (2004). *The SQUID handbook* **1**. Wiley-Vch.

[7] A.TH.A.M. de Waele and R. de Bruyn Ouboter (1969). "Quantum-interference phenomena in point contacts between two superconductors". *Physica* **41** (2): 225–254. Bibcode:1969Phy....41..225D. doi:10.1016/0031-8914(69)90116-5.

[8] Romani, G. L.; Williamson, S. J.; Kaufman, L. (1982). "Biomagnetic instrumentation". *Review of Scientific Instruments* **53** (12): 1815–1845. doi:10.1063/1.1136907. PMID 6760371.

[9] Sternickel, K.; Braginski, A. I. (2006). "Biomagnetism using SQUIDs: Status and perspectives". *Superconductor Science and Technology* **19** (3): S160. doi:10.1088/0953-2048/19/3/024.

[10] S.N. Erné, H.-D. Hahlbohm, H. Lübbig (1976). "Theory of the RF biased Superconducting Quantum Interference Device for the non-hysteretic regime". *J. Appl. Phys.* **47** (12): 5440–5442. Bibcode:1976JAP....47.5440E. doi:10.1063/1.322574.

[11] Cleuziou, J.-P.; Wernsdorfer, W. (2006). "Carbon nanotube superconducting quantum interference device". *nature nanotechnology* **1** (October). doi:10.1038/nnano.2006.54.

[12] Aprili, Marco (2006). "The nanoSQUID makes its debut". *nature nanotechnology* **1** (October).

[13] Kleiner, R.; Koelle, D.; Ludwig, F.; Clarke, J. (2004). "Superconducting quantum interference devices: State of the art and applications". *Proceedings of the IEEE* **92** (10): 1534. doi:10.1109/JPROC.2004.833655.

[14] microSQUID microscopy at Institut Néel (Grenoble, FRANCE)

[15] Clarke, J.; Lee, A.T.; Mück, M.; Richards, P.L. "Chapter 8.3: Nuclear Magnetic and Quadrupole Resonance and Magnetic Resonance Imaging". pp. 56–81. Missing or empty |title= (help) in Clarke & Braginski 2006

[16] Paik, Ho J. "Chapter 15.2: Superconducting Transducer for Gravitational-Wave Detectors". pp. 548–554. Missing or empty |title= (help) in Clarke & Braginski 2006

[17] "First Observation of the Dynamical Casimir Effect". Technology Review.

[18] Wilson, C. M. (2011). "Observation of the Dynamical Casimir Effect in a Superconducting Circuit". *Nature* **479** (7373): 376–379. arXiv:1105.4714. Bibcode:2011Natur.479..376W. doi:10.1038/nature10561. PMID 22094697.

[19] Quantum coherence with a single Cooper pair, V Bouchiat, D Vion, P Joyez, D Esteve, M H Devoret, 1998 Phys. Scr. 1998 165

[20] Ouellette, Jennifer. "SQUID Sensors Penetrate New Markets" (PDF). The Industrial Physicist. p. 22.

13.5   References

- Clarke, John; Braginski, Alex I., eds. (2006). *The SQUID Handbook: Applications of SQUIDs and SQUID Systems* **2**. Wiley-VCH. ISBN 978-3-527-40408-7.

Chapter 14

MiniGrail

MiniGRAIL is a type of Resonant Mass Antenna,[1] which is a massive sphere that used to detect gravitational waves. The MiniGRAIL is the first such detector to use a spherical design. It is located at Leiden University in the Netherlands. The project is being managed by the Kamerlingh Onnes Laboratory.[2] A team from the Department of Theoretical Physics of the University of Geneva, Switzerland, is also heavily involved.

Gravitational waves are a type of radiation that is emitted by objects that have mass and are undergoing acceleration. The strongest sources of gravitational waves are expected to be compact objects such as neutron stars and black holes. This detector may be able to detect certain types of instabilities in rotating single and binary neutron stars, and the merger of small black holes or neutron stars.[3]

A spherical design has the benefit of being able to detect gravitational waves arriving from any direction, and it is sensitive to polarization.[4] When gravitation waves with frequencies around 3,000 Hz pass through the MiniGRAIL ball, it will vibrate with displacements on the order of 10^{-20} m.[5] For comparison, the cross-section of a single proton (the nucleus of a hydrogen atom), is 10^{-15} m (1 fm).[6]

To improve sensitivity, the detector was intended to operate at a temperature of 20 mK.[2] The original antenna for the MiniGRAIL detector was a 68 cm diameter sphere made of an alloy of copper with 6% aluminium. This sphere had a mass of 1,150 kg and resonated at a frequency of 3,250 Hz. It was isolated from vibration by seven 140 kg masses. The bandwidth of the detector was expected to be ±230 Hz.[3]

During the casting of the sphere, a crack appeared that reduced the quality to unacceptable levels. It was replaced by a 68 cm sphere with a mass of 1,300 kg. This was manufactured by ItalBronze in Brazil. The larger mass lowered the resonant frequencies by about 200 Hz.[7] The sphere is suspended from stainless steel cables to which springs and masses are attached to dampen vibrations. Cooling is accomplished using a dilution refrigerator.[8]

Tests at temperatures of 5 K showed the detector to have a peak strain sensitivity of 1.5×10^{-20} Hz$^{-\frac{1}{2}}$ at a frequency of 2942.9 Hz. Over a bandwidth of 30 Hz, the strain sensitivity was more than 5×10^{-20} Hz$^{-\frac{1}{2}}$. This sensitivity is expected to improve by an order of magnitude when the instrument is operating at 50 mK.[4]

A similar detector named "Mario Schenberg" is being built in São Paulo, which will strongly increase the chances of detection by looking at coincidences.[9]

14.1   References

[1] Schutz , Bernard. *A First Course in General Relativity* (PDF) (2nd ed.). Cambridge. pp. 214–220. ISBN 978-0521887052.

[2] de Waard, A; et al. (2003). "MiniGRAIL, the first spherical detector". *Classical and Quantum Gravity* **20** (10): S143–S151. Bibcode:2003CQGra..20S.143D. doi:10.1088/0264-9381/20/10/317.

[3] Van Houwelingen, Jeroen (2002-06-24). "Development of a superconducting thin-film Nb-coil for use in the MiniGRAIL transducers" (PDF). Leiden University. pp. 1–17. Retrieved 2009-09-16.

[4] Gottardi, L.; De Waard, A.; Usenko, O.; Frossati, G.; Podt, M.; Flokstra, J.; Bassan, M.; Fafone, V.; et al. (November 2007). "Sensitivity of the spherical gravitational wave detector MiniGRAIL operating at 5K". *Physical Review D* **76** (10): 102005.1–102005.10. arXiv:0705.0122. Bibcode:2007PhRvD..76j2005G. doi:10.1103/PhysRevD.76.102005.

[5] Bruins, Eppo (2004-11-26). "'Listen, two black holes are clashing!'". innovations-report. Retrieved 2009-09-16.

[6] Ford, Kenneth William (2005). *The quantum world: quantum physics for everyone.* Harvard University Press. p. 11. ISBN 0-674-01832-X.

[7] de Waard, A.; et al. (2005). "MiniGRAIL progress report 2004". *Classical and Quantum Gravity* **22** (10): S215–S219. Bibcode:2005CQGra..22S.215D. doi:10.1088/0264-9381/22/10/012.

[8] de Waard, A.; et al. (March 2004). "Cooling down MiniGRAIL to milli-Kelvin temperatures". *Classical and Quantum Gravity* **21** (5): S465–S471. Bibcode:2004CQGra..21S.465D. doi:10.1088/0264-9381/21/5/012.

[9] Frajuca, Carlos; et al. (December 2005). "Resonant transducers for spherical gravitational wave detectors". *Brazilian Journal of Physics* **35** (4b): 1201–1203. Bibcode:2005BrJPh..35.1201F. doi:10.1590/S0103-97332005000700050.

14.2 External links

- MiniGRAIL on the internet

Chapter 15

GEO600

GEO600 is a gravitational wave detector located near Sarstedt in the South of Hanover, Germany. This instrument, and its sister interferometric detectors, when operational, are some of the most sensitive gravitational wave detectors ever designed. They are designed to detect relative changes in distance of the order of 10^{-21}, about the size of a single atom compared to the distance from the Sun to the Earth. GEO600 is capable of detecting gravitational waves in the frequency range 50 Hz to 1.5 kHz.[1] Construction on the project began in 1995.[2]

15.1 Hardware

GEO600 is a Michelson interferometer. It consists of two 600 meter long arms, which the laser beam passes twice, so that the effective optical arm length is 1200 m. The major optical components are located in an ultra-high vacuum system. The pressure is in the range of 10^{-8} mbar.[1]

15.1.1 Suspensions and seismic isolation

For precise measurements, the optics must be isolated from ground motion and other influences from the environment. For this reason, all ground-based interferometric gravitational wave detectors suspend their mirrors as multi-stage pendulums. For frequencies above the pendulum resonance frequency, pendulums provide a good isolation against vibrations. All the main optics of GEO600 are suspended as triple pendulums, to isolate the mirrors from vibrations in the horizontal plane. The uppermost and the intermediate mass are hung from cantilever springs, which provide isolation against vertical movement. On the uppermost mass are six coil- magnet actuators that are used to actively dampen the pendulums.[3] Furthermore, the whole suspension cage sits on piezo crystals. The crystals are used for an 'active seismic isolation system'. It moves the whole suspension in the opposite direction of the ground motion, so that ground motion is cancelled.[4]

15.1.2 Optics

The main mirrors of GEO600 are cylinders of fused silica with a diameter of 18 cm and a height of 10 cm. The beam splitter (with dimensions of 26 cm diameter and 8 cm thickness) is the only transmissive piece of optics in the high power path, therefore it was made from special grade fused silica. Its absorption has been measured to be smaller than 0.25 ppm/cm.[5]

15.1.3 Advanced features

GEO600 uses many advanced techniques and hardware that are planned to be used in the next generation of ground based gravitational wave detectors:

- Signal recycling: An additional mirror at the output of the interferometer forms a resonant cavity together with the end mirrors and thus increases a potential signal.

- Monolithic suspensions: The mirrors are suspended as pendulums. While steel wires are used for secondary mirrors, GEO's main mirrors are hanging from so called 'monolithic' suspensions. This means that the wires are made from the same material as the mirror: fused silica. The reason is that fused silica has less mechanical losses, and losses lead to noise.

- Electrostatic drives: Actuators are needed to keep the mirrors in their position and to align them. Secondary mirrors of GEO600 have magnets glued to them for this purpose. They can then be moved by coils. Since gluing magnets to mirrors will increase mechanical losses, the main mirrors of GEO600 use electrostatic drives (ESDs). The ESDs are a comb-like structure of electrodes at the back side of the mirror. If a voltage is applied to the electrodes, they produce an inhomogeneous electric field. The mirror will feel a force in this field.

- Thermal mirror actuation system: A circular heater is sitting behind the far east mirror. Due to inhomogeneous thermal expansion the radius of curvature of the mirror changes. The heater allows thermal tuning of the mirror's radius of curvature.[6]

- Output Mode Cleaner (OMC): An additional cavity at the output of the interferometer in front of the photodiode. Its purpose is to filter out light that does not potentially carry a gravitational wave signal.[7]

- Homodyne detection (also called 'DC readout') [8]

- Squeezing: Squeezed vacuum is injected into the dark port of the beam splitter. The use of squeezing can improve the sensitivity of GEO600 above 700 Hz by a factor of 1.5.[9]

A further difference to other projects is that GEO600 has no arm cavities.

15.2 Sensitivity and measurements

The sensitivity for gravitational wave strain is usually measured in amplitude spectral density (ASD). The peak sensitivity of GEO600 in this unit is $2{\times}10^{-22}$ 1/$\sqrt{\text{Hz}}$ at 600 Hz.[10] At high frequencies the sensitivity is limited by the available laser power. At the low frequency end, the sensitivity of GEO600 is limited by seismic ground motion.

15.3 Data/ Einstein@home

Not only the output of the main photodiode is registered, but also the output of a number of secondary sensors, for example photodiodes that measure auxiliary laser beams, microphones, seismometers, accelerometers, magnetometers and the performance of all the control circuits. These secondary sensors are important for diagnosis and to detect environmental influences on the interferometer output. The data stream is partly analyzed by the distributed computing project 'Einstein@home', software that volunteers can run on their computer.

From September 2011, both VIRGO and the LIGO detectors will be shut down for upgrades, leaving GEO600 as the only operating large scale laser interferometer searching for gravitational waves.[11]

15.4 Joint science run with LIGO

In November 2005, it was announced that the LIGO and GEO instruments have begun an extended joint *science run*. The three instruments (LIGO's instruments are located near Livingston, Louisiana and on the Hanford Site, Washington in the U.S.) will collect data for more than a year, with breaks for tuning and updates. This will be the fifth science run of GEO600. No signals were detected on previous runs, but the sensitivity of the instruments (and the quality of the data analysis) is continually improving, and once the data from the current run are analyzed, it is hoped that they will perhaps reveal the arrival at Earth of two unambiguous bursts of gravitational waves. This would constitute the first direct detection of gravitational radiation.

15.5 Claimed link between GEO600 detector noise and holographic properties of spacetime

On January 15, 2009, it was reported in *New Scientist* that some yet unidentified noise that was present in the GEO600 detector measurements might be because the instrument is sensitive to extremely small quantum fluctuations of space-time affecting the positions of parts of the detector.[12] This claim was made by Craig Hogan, a scientist from Fermilab, on the basis of his own theory of how such fluctuations should occur motivated by the holographic principle.[13]

The New Scientist story states that Hogan sent his prediction of "holographic noise" to the GEO600 collaboration in June 2008, and subsequently received a plot of the excess noise which "looked exactly the same as my prediction". However, Hogan knew before that time that the experiment was finding excess noise. Hogan's article published in *Physical Review D* in May 2008 states: "The approximate agreement of predicted holographic noise with otherwise unexplained noise in GEO600 motivates further study."[14] Hogan cites a 2007 talk from the GEO600 collaboration which already mentions "mid-band 'mystery' noise", and where the noise spectra are plotted.[15] A similar remark was made ("In the region between 100 Hz and 500 Hz a discrepancy between the uncorrelated sum of all noise projections and the actual observed sensitivity is found.") in a GEO600 paper submitted in October 2007 and published in May 2008.[16]

It is also a very common occurrence for gravitational wave detectors to find excess noise that is subsequently eliminated. According to Karsten Danzmann, the GEO600 principal investigator, "The daily business of improving the sensitivity of these experiments always throws up some excess noise (...). We work to identify its cause, get rid of it and tackle the next source of excess noise."[12] Additionally, some new estimates of the level of holographic noise in interferometry show that it must be much smaller in magnitude than was claimed by Hogan.[17]

15.6 See also

- Gravitational radiation

- Fermilab Holometer

- LIGO, for the two American gravitational interferometric detectors.

- eLISA, for the space-based ESA gravitational wave detector

- VIRGO, for the European gravitational interferometric detector.

- TAMA 300, for a Japanese gravitational interferometric detector.

- Einstein@Home, for a volunteer distributed computing program one can download in order to help the LIGO/GEO teams analyze their data

15.7 References

[1] "GEO600 Specifications". 2007. Retrieved 2007-06-26.

[2] http://www.geo600.de/general-information/history-purpose/

[3] Gossler, Stefan; et al. (2002). "The modecleaner system and suspension aspects of GEO600". *Class. Quantum Grav.* **19** (7): 1835. Bibcode:2002CQGra..19.1835G. doi:10.1088/0264-9381/19/7/382.

[4] Plissi, M.V.; et al. (2000). "GEO600 triple pendulum suspension system: Seismic isolation and control". *Rev. Sci. Instrum.* **71** (6): 2539. Bibcode:2000RScI...71.2539P. doi:10.1063/1.1150645.

[5] Hild, Stefan; et al. (2006). "Measurement of a low-absorption sample of OH-reduced fused silica". *Applied Optics* **45** (28): 7269. Bibcode:2006ApOpt..45.7269H. doi:10.1364/AO.45.007269.

[6] Lueck, H; et al. (2004). "Thermal correction of the radii of curvature of mirrors for GEO600". *Class. Quantum Grav.* **21** (5). Bibcode:2004CQGra..21S.985L. doi:10.1088/0264-9381/21/5/090.

[7] Prijatelj, Miro; et al. (2012). "The output mode cleaner of GEO600". *Class. Quantum Grav.* **29** (5). Bibcode:2012CQGra..29e5009P. doi:10.1088/0264-9381/29/5/055009.

[8] Hild, Stefan; et al. (2009). "DC-readout of a signal-recycled gravitational wave detector". *Class. Quantum Grav.* **26** (5). arXiv:0811.3242. Bibcode:2009CQGra..26e5012H. doi:10.1088/0264-9381/26/5/055012.

[9] The LIGO scientific collaboration (2011). "A gravitational wave observatory operating beyond the quantum shot-noise limit". *Nature Physics* **7** (12). doi:10.1038/nphys2083.

[10] "GEO600 Sensitivity". Retrieved 2013-05-17.

[11] "GWIC roadmap p.65" (PDF). Retrieved 2013-05-17.

[12] New Scientist - Our world may be a giant hologram

[13] Hogan, Craig J.; Mark G. Jackson (June 2009). "Holographic geometry and noise in matrix theory". *Phys. Rev. D* **79** (12): 124009. arXiv:0812.1285. Bibcode:2009PhRvD..79l4009H. doi:10.1103/PhysRevD.79.124009.

[14] Hogan, Craig J. (2008). "Measurement of quantum fluctuations in geometry". *Phys. Rev. D* **77** (10): 104031. arXiv:0712.3419. Bibcode:2008PhRvD..77j4031H. doi:10.1103/PhysRevD.77.104031.

[15] http://www.ligo.caltech.edu/docs/G/G070506-00.pdf Talk by K. Strain "The Status of GEO600"

[16] http://www.iop.org/EJ/abstract/0264-9381/25/11/114043 GEO600 paper mentioning unexplained noise in 2007

[17] Smolyaninov, Igor I. (Apr 2009). "Level of holographic noise in interferometry". *Phys. Rev. D* **78** (8): 087503. arXiv:0903.4129. Bibcode:2009PhRvD..79h7503S. doi:10.1103/PhysRevD.79.087503.

15.8   External links

- GEO600 home page, the official website of the GEO600 project.

- Cardiff Gravity Group, a page describing research at Cardiff University in Wales, including collaboration in the GEO 600 project, includes an excellent list of tutorials on gravitational wave radiation.

- Amos, Jonathan. Science to ride gravitational waves. November 8, 2005. *BBC News*.

Coordinates: 52°14′49″N 9°48′30″E / 52.24694°N 9.80833°E

Chapter 16

TAMA 300

Vacuum pump and the laser beam duct of TAMA300

TAMA 300 was a gravitational wave detector located at the Mitaka campus of the National Astronomical Observatory of Japan. It is a project of the gravitational wave studies group at the Institute for Cosmic Ray Research (ICRR) of the University of Tokyo. The ICRR was established in 1976 for cosmic ray studies, and is currently developing the Large Scale Cryogenic Gravitational Wave Telescope (LCGT).

The TAMA project started in 1995 and stopped data taking in 2003. It adopts a Fabry Perot Michelson Interferometer (FPMI) with power recycling. It is officially known as the **300m Laser Interferometer Gravitational Wave Antenna**.

The goal of the project is to develop advanced techniques needed for a future kilometer sized interferometer and to detect gravitational waves that may occur by chance within our local group of galaxies.

16.1 External links

- Official website

Coordinates: 35°40′36″N 139°32′10″E / 35.67667°N 139.53611°E

Chapter 17

KAGRA

The **Kamioka Gravitational Wave Detector (KAGRA)**, formerly the **Large Scale Cryogenic Gravitational Wave Telescope (LCGT)**, is a future project of the gravitational wave studies group at the Institute for Cosmic Ray Research (ICRR) of the University of Tokyo. The ICRR was established in 1976 for cosmic ray studies, and is currently working on TAMA 300. The LCGT project was approved on 22 June 2010. In January 2012, it was given its new name, KAGRA, deriving the "KA" from its location at the Kamioka mine and "GRA" from gravity and gravitational radiation.[1]

KAGRA has two sets of 3 km (1.9 mi) arm length laser interferometric gravitational wave detectors which were being built in tunnels of Kamioka mine in Japan. The excavation phase of tunnels was completed on 31 March 2014. KAGRA will detect chirp waves from binary neutron star coalescence at 240 Mpc away with a S/N of 10. The expected number of detectable events in a year is two or three. To achieve the required sensitivity, several advanced techniques will be employed such as a low-frequency vibration-isolation system, a suspension point interferometer, cryogenic mirrors, a resonant side band extraction method, a high-power laser system and so on. KAGRA was initially hoped to begin operations in 2009[2] but is now likely to enter operation in 2018.[3]

17.1 References

[1] "LCGT got new nickname "KAGRA"".

[2] Uchiyama, T.; et al. (2004). "Present status of large-scale cryogenic gravitational wave telescope". *Class. Quantum Grav.* **21** (5): S1161–S1172. Bibcode:2004CQGra..21S1161U. doi:10.1088/0264-9381/21/5/115., free version available at "Present status of large-scale cryogenic gravitational wave telescope" (PDF).

[3] Kuroda, K; et al. (April 2010). "Status of LCGT". *Class. Quantum Grav.* **27** (8): 084004. Bibcode:2010CQGra..27h4004K. doi:10.1088/0264-9381/27/8/084004., free version available at "Status of LCGT" (PDF).

17.2 External

- Official page (English)

Chapter 18

Deci-hertz Interferometer Gravitational wave Observatory

The **DECI-Hertz Interferometer Gravitational wave Observatory** (or **DECIGO**) is a proposed Japanese, space-based, gravitational wave observatory.[1] The laser interferometric gravitational wave detector is so named because it is to be most sensitive in the frequency band between 0.1 Hz and 10 Hz (a decihertz).[2] It is anticipated for launch in 2027.

18.1　See also

- List of proposed space observatories

18.2　References

[1] "The Japanese space gravitational wave antenna - DECIGO". *J. Phys.: Conf. Ser.* **122** (1). 2008. Bibcode:2008JPhCS.122a2006K. doi:10.1088/1742-6596/122/1/012006. |first1= missing |last1= in Authors list (help)

[2] "DECIGO: The Japanese space gravitational wave antenna". *J. Phys.: Conf. Ser.* **154** (1). 2009. Bibcode:2009JPhCS.154a2040S. doi:10.1088/1742-6596/154/1/012040. |first1= missing |last1= in Authors list (help)

Chapter 19

pp-wave spacetime

In general relativity, the **pp-wave spacetimes**, or **pp-waves** for short, are an important family of exact solutions of Einstein's field equation. These solutions model radiation moving at the speed of light. This radiation may consist of:

- electromagnetic radiation,

- gravitational radiation,

- *massless* radiation associated with some hypothetical distinct type relativistic classical field,

or any combination of these, so long as the radiation is all moving in the *same* direction.

A special type of pp-wave spacetime, the plane wave spacetimes, provide the most general analog in general relativity of the plane waves familiar to students of electromagnetism. In particular, in general relativity, we must take into account the gravitational effects of the energy density of the electromagnetic field itself. When we do this, *purely electromagnetic plane waves* provide the direct generalization of ordinary plane wave solutions in Maxwell's theory.

Furthermore, in general relativity, disturbances in the gravitational field itself can propagate, at the speed of light, as "wrinkles" in the curvature of spacetime. Such *gravitational radiation* is the gravitational field analog of electromagnetic radiation. In general relativity, the gravitational analogue of electromagnetic plane waves are precisely the vacuum solutions among the plane wave spacetimes. They are called gravitational plane waves.

There are physically important examples of pp-wave spacetimes which are *not* plane wave spacetimes. In particular, the physical experience of an observer who whizzes by a gravitating object (such as a star or a black hole) at nearly the speed of light can be modelled by an *impulsive* pp-wave spacetime called the Aichelburg–Sexl ultraboost. The gravitational field of a beam of light is modelled, in general relativity, by a certain axi-symmetric pp-wave.

Pp-waves were introduced by Hans Brinkmann in 1925 and have been rediscovered many times since, most notably by Albert Einstein and Nathan Rosen in 1937. The term *pp* stands for *plane-fronted waves with parallel propagation*, and was introduced in 1962 by Jürgen Ehlers and Wolfgang Kundt.

19.1   Mathematical definition

A *pp-wave spacetime* is any Lorentzian manifold whose metric tensor can be described, with respect to Brinkmann coordinates, in the form

$$ds^2 = H(u, x, y) \, du^2 + 2 \, du \, dv + dx^2 + dy^2$$

where H is any smooth function. This was the original definition of Brinkmann, and it has the virtue of being easy to understand.

The definition which is now standard in the literature is more sophisticated. It makes no reference to any coordinate chart, so it is a coordinate-free definition. It states that any Lorentzian manifold which admits a *covariantly constant* null vector field k is called a pp-wave spacetime. That is, the covariant derivative of k must vanish identically:

$$\nabla k = 0.$$

This definition was introduced by Ehlers and Kundt in 1962. To relate Brinkmann's definition to this one, take $k = \partial_v$, the coordinate vector orthogonal to the hypersurfaces $v = v_0$. In the *index-gymnastics* notation for tensor equations, the condition on k can be written $k_{a;b} = 0$.

Neither of these definitions make any mention of any field equation; in fact, they are *entirely independent of physics*. In this sense, the notion of a pp-wave spacetime is entirely mathematical and belongs to the study of pseudo-Riemannian geometry. In the next section, we will turn to the *physical interpretation* of pp-waves.

Ehlers and Kundt gave several more coordinate-free characterizations, including:

- A Lorentzian manifold is a pp-wave if and only if it admits a one-parameter subgroup of isometries having null orbits, and whose curvature tensor has vanishing eigenvalues.

- A Lorentzian manifold with nonvanishing curvature is a (nontrivial) pp-wave if and only if it admits a covariantly constant bivector. (If so, this bivector is a null bivector.)

19.2 Physical interpretation

It is a purely mathematical fact that the characteristic polynomial of the Einstein tensor of any pp-wave spacetime vanishes identically. Equivalently, we can find a Newman–Penrose complex null tetrad such that the Ricci-NP scalars Φ_{ij} (describing any matter or nongravitational fields which may be present in a spacetime) and the Weyl-NP scalars Ψ_i (describing any gravitational field which may be present) each have only one nonvanishing component. Specifically, with respect to the NP tetrad

$$\vec{\ell} = \partial_u - H/2 \, \partial_v$$

$$\vec{n} = \partial_v$$

$$\vec{m} = \frac{1}{\sqrt{2}} \left(\partial_x + i \, \partial_y \right)$$

the only nonvanishing component of the Ricci spinor is

$$\Phi_{00} = \frac{1}{4} \left(H_{xx} + H_{yy} \right)$$

and the only nonvanishing component of the Weyl spinor is

$$\Psi_0 = \frac{1}{4} \left((H_{xx} - H_{yy}) + 2i \, H_{xy} \right).$$

This means that any pp-wave spacetime can be interpreted, in the context of general relativity, as a null dust solution. Also, the Weyl tensor always has Petrov type **N** as may be verified by using the Bel criteria.

In other words, pp-waves model various kinds of *classical* and *massless* radiation traveling at the local speed of light. This radiation can be gravitational, electromagnetic, some hypothetical kind of massless radiation other than these two, or any combination of these. All this radiation is traveling in the same direction, and the null vector $k = \partial_v$ plays the role of a wave vector.

19.3 Relation to other classes of exact solutions

Unfortunately, the terminology concerning pp-waves, while fairly standard, is highly confusing and tends to promote misunderstanding.

In any pp-wave spacetime, the covariantly constant vector field k always has identically vanishing optical scalars. Therefore, pp-waves belong to the Kundt class (the class of Lorentzian manifolds admitting a null congruence with vanishing optical scalars).

Going in the other direction, pp-waves include several important special cases.

From the form of Ricci spinor given in the preceding section, it is immediately apparent that a pp-wave spacetime (written in the Brinkmann chart) is a vacuum solution if and only if H is a harmonic function (with respect to the spatial coordinates x, y). Physically, these represent purely gravitational radiation propagating along the null rays ∂_v .

Ehlers and Kundt and Sippel and Gönner have classified vacuum pp-wave spacetimes by their autometry group, or group of *self-isometries*. This is always a Lie group, and as usual it is easier to classify the underlying Lie algebras of Killing vector fields. It turns out that the most general pp-wave spacetime has only one Killing vector field, the null geodesic congruence $k = \partial_v$. However, for various special forms of H , there are additional Killing vector fields.

The most important class of particularly symmetric pp-waves are the plane wave spacetimes, which were first studied by Baldwin and Jeffery. A plane wave is a pp-wave in which H is quadratic, and can hence be transformed to the simple form

$$H(u, x, y) = a(u)\,(x^2 - y^2) + 2\,b(u)\,xy + c(u)\,(x^2 + y^2)$$

Here, a, b, c are arbitrary smooth functions of u . Physically speaking, a, b describe the wave profiles of the two linearly independent polarization modes of gravitational radiation which may be present, while c describes the wave profile of any nongravitational radiation. If $c = 0$, we have the vacuum plane waves, which are often called plane gravitational waves.

Equivalently, a plane-wave is a pp-wave with at least a five-dimensional Lie algebra of Killing vector fields X , including $X = \partial_v$ and four more which have the form

$$X = \frac{\partial}{\partial u}(px + qy)\,\partial_v + p\,\partial_x + q\,\partial_y$$

where

$$\ddot{p} = -ap + bq - cp$$

$$\ddot{q} = aq - bp - cq.$$

Intuitively, the distinction is that the wavefronts of plane waves are truly *planar*; all points on a given two-dimensional wavefront are equivalent. This not quite true for more general pp-waves. Plane waves are important for many reasons; to mention just one, they are essential for the beautiful topic of colliding plane waves.

A more general subclass consists of the **axisymmetric pp-waves**, which in general have a two-dimensional Abelian Lie algebra of Killing vector fields. These are also called *SG2 plane waves*, because they are the second type in the symmetry classification of Sippel and Gönner. A limiting case of certain axisymmetric pp-waves yields the Aichelburg/Sexl ultraboost modeling an ultrarelativistic encounter with an isolated spherically symmetric object.

(See also the article on plane wave spacetimes for a discussion of physically important special cases of plane waves.)

J. D. Steele has introduced the notion of **generalised pp-wave spacetimes**. These are nonflat Lorentzian spacetimes which admit a self-dual covariantly constant null bivector field. The name is potentially misleading, since as Steele points out, these are nominally a *special case* of nonflat pp-waves in the sense defined above. They are only a generalization in the sense that although the Brinkmann metric form is preserved, they are not necessarily the vacuum solutions studied by Ehlers and Kundt, Sippel and Gönner, etc.

Another important special class of pp-waves are the sandwich waves. These have vanishing curvature except on some range $u_1 < u < u_2$, and represent a gravitational wave moving through a Minkowski spacetime background.

19.4   Relation to other theories

Since they constitute a very simple and natural class of Lorentzian manifolds, defined in terms of a null congruence, it is not very surprising that they are also important in other relativistic classical field theories of gravitation. In particular, pp-waves are exact solutions in the Brans–Dicke theory, various higher curvature theories and Kaluza–Klein theories, and certain gravitation theories of J. W. Moffat. Indeed, B. O. J. Tupper has shown that the *common* vacuum solutions in general relativity and in the Brans/Dicke theory are precisely the vacuum pp-waves (but the Brans/Dicke theory admits further wavelike solutions). Hans-Jürgen Schmidt has reformulated the theory of (four-dimensional) pp-waves in terms of a *two-dimensional* **metric-dilaton** theory of gravity.

Pp-waves also play an important role in the search for quantum gravity, because as Gary Gibbons has pointed out, all loop term quantum corrections vanish identically for any pp-wave spacetime. This means that studying tree-level quantizations of pp-wave spacetimes offers a glimpse into the yet unknown world of quantum gravity.

It is natural to generalize pp-waves to higher dimensions, where they enjoy similar properties to those we have discussed. C. M. Hull has shown that such *higher-dimensional pp-waves* are essential building blocks for eleven-dimensional supergravity.

19.5   Geometric and physical properties

PP-waves enjoy numerous striking properties. Some of their more abstract mathematical properties have already been mentioned. In this section we can discuss only a few additional properties.

Consider an inertial observer in Minkowski spacetime who encounters a sandwich plane wave. Such an observer will experience some interesting optical effects. If he looks into the *oncoming* wavefronts at distant galaxies which have already encountered the wave, he will see their images undistorted. This must be the case, since he cannot know the wave is coming until it reaches his location, for it is traveling at the speed of light. However, this can be confirmed by direct computation of the optical scalars of the null congruence ∂_v . Now suppose that after the wave passes, our observer turns about face and looks through the *departing* wavefronts at distant galaxies which the wave has not yet reached. Now he sees their optical images sheared and magnified (or demagnified) in a time-dependent manner. If the wave happens to be a polarized *gravitational plane wave*, he will see circular images alternately squeezed horizontally while expanded vertically, and squeezed vertically while expanded horizontally. This directly exhibits the characteristic effect of a gravitational wave in general relativity on light.

The effect of a passing polarized gravitational plane wave on the relative positions of a cloud of (initially static) test particles will be qualitatively very similar. We might mention here that in general, the motion of test particles in pp-wave spacetimes can exhibit chaos.

The fact that Einstein's field equation is nonlinear is well-known. This implies that if you have two exact solutions, there is almost never any way to linearly superimpose them. PP waves provide a rare exception to this rule: if you have two PP waves sharing the same covariantly constant null vector (the same geodesic null congruence, i.e. the same wave vector field), with metric functions H_1, H_2 respectively, then $H_1 + H_2$ gives a third exact solution.

Roger Penrose has observed that near a null geodesic, *every Lorentzian spacetime looks like a plane wave*. To show this, he used techniques imported from algebraic geometry to "blow up" the spacetime so that the given null geodesic becomes the covariantly constant null geodesic congruence of a plane wave. This construction is called a Penrose limit.

Penrose also pointed out that in a pp-wave spacetime, all the polynomial scalar invariants of the Riemann tensor *vanish identically*, yet the curvature is almost never zero. This is because in four-dimension all pp-waves belong to the class of VSI spacetimes. Such statement does not hold in higher-dimensions since there are higher-dimensional pp-waves of algebraic type II with non-vanishing polynomial scalar invariants. If you view the Riemann tensor as a second rank tensor acting on bivectors, the vanishing of invariants is analogous to the fact that a nonzero null vector has vanishing squared length.

Penrose was also the first to understand the strange nature of causality in pp-sandwich wave spacetimes. He showed that some or all of the null geodesics emitted at a given event will be refocused at a later event (or string of events). The details depend upon whether the wave is purely gravitational, purely electromagnetic, or neither.

Every pp-wave admits many different Brinkmann charts. These are related by coordinate transformations, which in this context may be considered to be gauge transformations. In the case of plane waves, these gauge transformations allow us to always regard two colliding plane waves to have *parallel wavefronts*, and thus the waves can be said to *collide head-on*. This is an exact result in fully nonlinear general relativity which is analogous to a similar result concerning electromagnetic plane waves as treated in special relativity.

19.6 Examples

There are many noteworthy *explicit* examples of pp-waves. ("Explicit" means that the metric functions can be written down in terms of elementary functions or perhaps well-known special functions such as Mathieu functions.)

Explicit examples of *axisymmetric pp-waves* include

- The Aichelburg–Sexl ultraboost is an impulsive plane wave which models the physical experience of an observer who whizzes by a spherically symmetric gravitating object at nearly the speed of light,

- The Bonnor beam is an axisymmetric plane wave which models the gravitational field of an infinitely long beam of incoherent electromagnetic radiation.

Explicit examples of *plane wave spacetimes* include

- exact monochromatic gravitational plane wave and monochromatic electromagnetic plane wave solutions, which generalize solutions which are well-known from weak-field approximation,

- the Schwarzschild generating plane wave, a gravitational plane wave which, should it collide head-on with a twin, will produce in the *interaction zone* of the resulting colliding plane wave solution a region which is locally isometric to part of the *interior* of a Schwarzschild black hole, thereby permitting a classical peek at the local geometry *inside* the event horizon,

- the uniform electromagnetic plane wave; this spacetime is foliated by spacelike hyperslices which are isometric to S^3,

- the wave of death is a gravitational plane wave exhibiting a *strong nonscalar null curvature singularity*, which propagates through an initially flat spacetime, progressively destroying the universe,

- homogeneous plane waves, or *SG11 plane waves* (type 11 in the Sippel and Gönner symmetry classification), which exhibit a *weak nonscalar null curvature singularity* and which arise as the Penrose limits of an appropriate null geodesic approaching the curvature singularity which is present in many physically important solutions, including the Schwarzschild black holes and FRW cosmological models.

19.7 See also

- Gravitational wave

- Newman–Penrose formalism

19.8   References

- "On Generalised P.P. Waves" (PDF). *J. D. Steele*. Retrieved June 12, 2005.

- Hall, Graham (2004). *Symmetries and Curvature Structure in General Relativity (World Scientific Lecture Notes in Physics)*. Singapore: World Scientific Pub. Co. ISBN 981-02-1051-5.

- Stephani, Hans; Kramer, Dietrich; MacCallum, Malcolm; Hoenselaers, Cornelius & Herlt, Eduard (2003). *Exact Solutions of Einstein's Field Equations*. Cambridge: Cambridge University Press. ISBN 0-521-46136-7. *See Section 24.5*

- Sippel, R.; & Gönner, H. (1986). "Symmetry classes of pp waves". *Gen. Rel. Grav.* **12**: 1129–1243.

- Penrose, Roger (1976). "Any spacetime has a plane wave as a limit". *Differential Geometry and Relativity*. pp. 271–275.

- Tupper, B. O. J. (1974). "Common solutions of the Einstein and Brans-Dicke theories". *Int. J. Th. Phys.* **11** (5): 353–356. Bibcode:1974IJTP...11..353T. doi:10.1007/BF01808090.

- Penrose, Roger (1965). "A remarkable property of plane waves in general relativity". *Rev. Mod. Phys.* **37**: 215–220. Bibcode:1965RvMP...37..215P. doi:10.1103/RevModPhys.37.215.

- Ehlers, Jürgen; & Kundt, Wolfgang (1962). "Exact solutions of the gravitational field equations". *Gravitation: an Introduction to Current Research*. pp. 49–101. *See Section 2-5*

- Baldwin, O. R.; and Jeffery, G. B (1926). "The relativity theory of plane waves". *Proc. Roy. Soc. Lond.* A **111** (757): 95. Bibcode:1926RSPSA.111...95B. doi:10.1098/rspa.1926.0051.

- H. W. Brinkmann (1925). "Einstein spaces which are mapped conformally on each other". *Math. Ann.* **18**: 119. doi:10.1007/BF01208647.

- Yi-Fei Chen and J.X. Lu (2004)"Generating a dynamical M2 brane from super-gravitons in a pp-wave background"

- Bum-Hoon Lee(2005)"D-branes in the pp-wave background"

- H.-J. Schmidt (1998). "arXiv:gr-qc/9712034: A two-dimensional representation of four-dimensional gravitational waves, Int. J. Mod. Phys. D7 (1998) 215-224"

19.9   External links

- Pp-wave on arxiv.org

Chapter 20

Spin-flip

This article is about black hole spin-flips. For atomic spin-flips, see Hydrogen line.
  A **black hole spin-flip** occurs when the spin axis of a rotating black hole undergoes a sudden change in orientation due

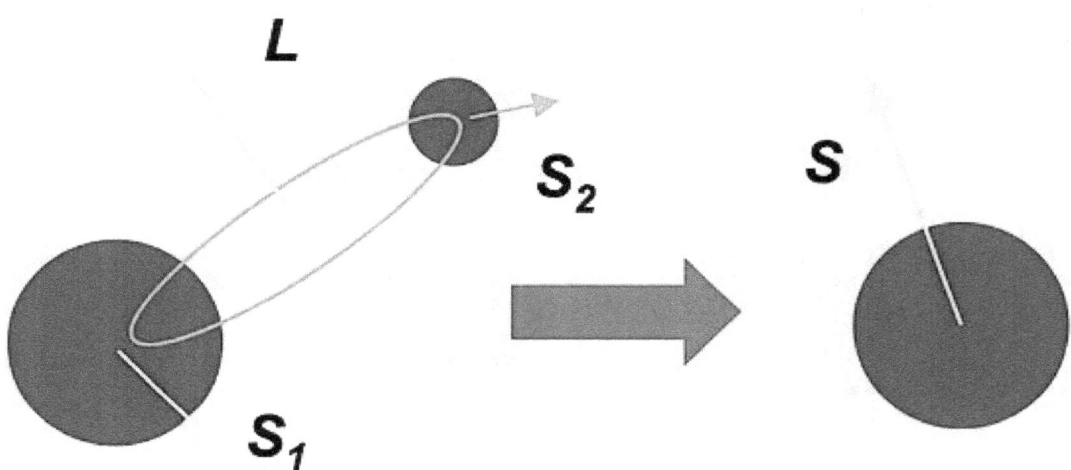

Schematic diagram of a black hole spin-flip.

to absorption of a second (smaller) black hole. Spin-flips are believed to be a consequence of galaxy mergers, when two supermassive black holes form a bound pair at the center of the merged galaxy and coalesce after emitting gravitational waves. Spin-flips are significant astrophysically since a number of physical processes are associated with black hole spins; for instance, jets in active galaxies are believed to be launched parallel to the spin axes of supermassive black holes. A change in the rotation axis of a black hole due to a spin-flip would therefore result in a change in the direction of the jet.

20.1   Physics of Spin-Flips

A spin-flip is a late stage in the evolution of a binary black hole. The binary consists of two black holes, with masses M_1 and M_2, that revolve around their common center of mass. The total angular momentum J of the binary system is the sum of the angular momentum of the orbit, L, plus the spin angular momenta $S_{1,2} = S_1 + S_2$ of the two holes. If we write $\mathbf{M_1}, \mathbf{M_2}$ as the masses of each hole and $\mathbf{a_1}, \mathbf{a_2}$ as their Kerr parameters,[1] then use the angle from north of their spin axes as given by θ, we can write,

$\mathbf{S}_1 = \{\mathbf{a}_1 * \mathbf{M}_1 * \cos(\pi/2 - \theta), \mathbf{a}_1 * \mathbf{M}_1 * \sin(\pi/2 - \theta)\}$

$\mathbf{S}_2 = \{\mathbf{a}_2 * \mathbf{M}_2 * \cos(\pi/2 - \theta), \mathbf{a}_2 * \mathbf{M}_2 * \sin(\pi/2 - \theta)\}$

$\mathbf{J}_{\text{init}} = \mathbf{L}_{\text{orb}} + \mathbf{S}_1 + \mathbf{S}_2.$

If the orbital separation is sufficiently small, emission of energy and angular momentum in the form of gravitational radiation will cause the orbital separation to drop. Eventually, the smaller hole M_2 reaches the innermost stable circular orbit, or ISCO, around the larger hole. Once the ISCO is reached, there no longer exists a stable orbit, and the smaller hole plunges into the larger hole, coalescing with it. The final angular momentum after coalescence is just

$\mathbf{J}_{\text{final}} = \mathbf{S},$

the spin angular momentum of the single, coalesced hole. Neglecting the angular momentum that is carried away by gravitational waves during the final plunge—which is small[2]—conservation of angular momentum implies

$\mathbf{S} \approx \mathbf{L}_{\text{ISCO}} + \mathbf{S}_1 + \mathbf{S}_2.$

S_2 is of order $(M_2/M_1)^2$ times S_1 and can be ignored if M_2 is much smaller than M_1 . Making this approximation,

$\mathbf{S} \approx \mathbf{L}_{\text{ISCO}} + \mathbf{S}_1.$

This equation states that the final spin of the hole is the sum of the larger hole's initial spin plus the orbital angular momentum of the smaller hole at the last stable orbit. Since the vectors S_1 and L are generically oriented in different directions, S will point in a different direction than S_1 —a spin-flip.[3]

The angle by which the black hole's spin re-orients itself depends on the relative size of L_{ISCO} and S_1 , and on the angle between them. At one extreme, if S_1 is very small, the final spin will be dominated by L_{ISCO} and the flip angle can be large. At the other extreme, the larger black hole might be a maximally-rotating Kerr black hole initially. Its spin angular momentum is then of order

$S_1 \approx GM_1^2/c.$

The orbital angular momentum of the smaller hole at the ISCO depends on the direction of its orbit, but is of order

$L_{\text{ISCO}} \approx GM_1 M_2/c.$

Comparing these two expressions, it follows that even a fairly small hole, with mass about one-fifth that of the larger hole, can reorient the larger hole by 90 degrees or more.[3]

20.2   Connection with radio galaxies

Black hole spin-flips were first discussed[3] in the context of a particular class of radio galaxy, the X-shaped radio sources. The X-shaped galaxies exhibit two, misaligned pairs of radio lobes: the "active" lobes and the "wings". It is believed that the wings are oriented in the direction of the jet prior to the spin-flip, and that the active lobes point in the current jet direction. The spin-flip could have been caused by absorption of a second black hole during a galaxy merger.

20.3   See also

- Supermassive black hole

- Gravitational waves

- Radio galaxy

20.4   References

[1] Rosalba, Perna. KERR (SPINNING) BLACK HOLES [PowerPoint slides]. Retrieved from http://www.astro.sunysb.edu/rosalba/astro2030/KerrBH.pdf

[2] Baker, J. et al. (2006), "Gravitational wave extraction from an inspiraling configuration of merging black holes", *Phys. Rev. Lett.* **96**, 11102.

[3] D. Merritt and R. Ekers (2002), "Tracing black hole mergers through radio lobe morphology", *Science* **297**, 1310.

20.5 External links

- Massive black hole binary evolution An article on binary black holes.

- Scientists Detect "Smoking Gun" of Colliding Black Holes

Chapter 21

Sticky bead argument

In general relativity, the **sticky bead argument** is a simple thought experiment designed to show that gravitational radiation is indeed predicted by general relativity, and can have physical effects. These claims were not widely accepted prior to about 1955, but after the introduction of the bead argument, any remaining doubts soon disappeared from the research literature.

The argument is often credited to Hermann Bondi, who popularized it, but it was apparently originally proposed anonymously by Richard Feynman.[1][2]

21.1 Description of the thought experiment

The thought experiment was first described by Feynman (under the pseudonym "Mr. Smith") in 1957, at a conference at Chapel Hill, North Carolina.[2][3] His insight was that a passing gravitational wave should, in principle, cause a bead which is free to slide along a stick to move back and forth, when the stick is held transversely to the wave's direction of propagation. The wave generates tidal forces about the midpoint of the stick. These produce alternating, longitudinal tensile and compressive stresses in the material of the stick; but the bead, being free to slide, moves along the stick in response to the tidal forces. If contact between the bead and stick is 'sticky', then heating of both parts will occur due to friction. This heating, said Feynman, showed that the wave did indeed impart energy to the bead and rod system, so it must indeed transport energy.

21.2 History of arguments on the properties of gravitational waves

21.2.1 Einstein's double reversal

The creator of the theory of general relativity, Albert Einstein, argued in 1916 that gravitational radiation should be produced, according to his theory, by any mass-energy configuration which has a time-varying quadrupole moment (or higher multipole moment). Using a linearized field equation (appropriate for the study of *weak* gravitational fields), he derived the famous quadrupole formula quantifying the rate at which such radiation should carry away energy. Examples of systems with time varying quadrupole moments include vibrating strings, bars rotating about an axis perpendicular to the symmetry axis of the bar, and binary star systems, but not rotating disks.

In 1922, Arthur Stanley Eddington wrote a paper expressing (apparently for the first time) the view that gravitational waves are in essence ripples in coordinates, and have no physical meaning. He did not appreciate Einstein's arguments that the waves are real.

In 1936, together with Nathan Rosen, Einstein rediscovered the Beck vacuums, a family of exact gravitational wave solutions with cylindrical symmetry (sometimes also called *Einstein-Rosen waves*). While investigating the motion of test particles in these solutions, Einstein and Rosen became convinced that gravitational waves were unstable to collapse.

Einstein reversed himself and declared that gravitational radiation was *not* after all a prediction of his theory. Einstein wrote to his friend Max Born

> Together with a young collaborator, I arrived at the interesting result that gravitational waves do not exist, though they had been assumed a certainty to the first approximation. This shows that the nonlinear field equations can show us more, or rather limit us more, than we have believed up till now.

In other words, Einstein believed that he and Rosen had established that their new argument showed that the prediction of gravitational radiation was a *mathematical artifact* of the linear approximation he had employed in 1916. Einstein believed these plane waves would gravitationally collapse into points; he had long hoped something like this would explain quantum mechanical wave-particle duality.

Einstein and Rosen accordingly submitted a paper entitled *Do gravitational waves exist?* to a leading physics journal, the Physical Review, in which they described their wave solutions and concluded that the "radiation" which seemed to appear in general relativity was not genuine radiation capable of transporting energy or having (in principle) measurable physical effects. The anonymous referee, who—as the current editor of the Physical Review recently confirmed, all parties now being deceased—was the combative cosmologist, Howard Percy Robertson, pointed out the error described below, and the manuscript was returned to the authors with a note from the editor asking them to revise the paper to address these concerns. Quite uncharacteristically, Einstein took this criticism very badly, angrily replying "I see no reason to address the, in any case erroneous, opinion expressed by your referee." He vowed never again to submit a paper to the Physical Review. Instead, Einstein and Rosen resubmitted the paper without change to another and much less well known journal, the Journal of the Franklin Institute. He kept his vow regarding the Physical Review.

Leopold Infeld, who arrived at Princeton University at this time, later remembered his utter astonishment on hearing of this development, since radiation is such an essential element for any classical field theory worthy of the name. Infeld expressed his doubts to a leading expert on general relativity: H. P. Robertson, who had just returned from a visit to Caltech. Going over the argument as Infeld remembered it (apparently from a conversation with Einstein), Robertson was able to show Infeld the mistake: locally, the Einstein-Rosen waves are gravitational plane waves (which had been studied earlier by O. R. Baldwin and George Barker Jeffery, and even earlier by Hans W. Brinkmann). Einstein and Rosen had correctly shown that a cloud of test particles would, in sinusoidal plane waves, form caustics, but changing to another chart (essentially the Brinkmann coordinates) shows that the formation of the caustic is *not a contradiction at all*, but in fact just what one would expect in this situation. Infeld then approached Einstein, who concurred with this analysis (still not knowing it was he who reviewed the Physical Review submission).

Since Rosen had recently departed for the Soviet Union, Einstein acted alone in promptly and thoroughly revising their joint paper. This third version was retitled *On gravitational waves*, and, following Robertson's suggestion of a transformation to cylindrical coordinates, presented what are now called Einstein-Rosen cylindrical waves (these are locally isometric to plane waves). This is the version which eventually appeared. However, Rosen was unhappy with this revision and eventually published his own version, which retained the erroneous "disproof" of the prediction of gravitational radiation.

In a letter to the editor of the Physical Review, Robertson wryly reported that in the end, Einstein had fully accepted the objections which had initially so upset him.

21.2.2   The Bern and Chapel Hill conferences

In 1955, an important conference honoring the semi-centennial of special relativity was held in Bern, the Swiss town where Einstein was working in the famous patent office during the Annus mirabilis. Rosen attended and gave a talk in which he computed the *Einstein pseudotensor* and *Landau-Lifschitz pseudotensor* (two alternative, non-covariant, descriptions of the energy carried by a *gravitational* field, a notion which is notoriously difficult to pin down in general relativity). These turn out to be zero for the Einstein-Rosen waves, and Rosen argued that this reaffirmed the negative conclusion he had reached with Einstein in 1936.

However, by this time a few physicists, such as Felix A. E. Pirani and Ivor Robinson, had come to appreciate the role played by curvature in producing tidal accelerations, and were able to convince many peers that gravitational radiation would indeed be produced, at least in cases such as a vibrating spring where different pieces of the system were clearly not in inertial motion. Nonetheless, some physicists continued to doubt whether radiation would be produced by a binary star system, where the world lines of the centers of mass of the two stars should, according to the EIH approximation (dating from 1938 and due to Einstein, Infeld, and Banesh Hoffmann), follow timelike geodesics.

Inspired by conversations by Felix Pirani, Hermann Bondi took up the study of gravitational radiation, in particular the question of quantifying the energy and momentum carried off 'to infinity' by a radiating system. During the next few years, Bondi developed the Bondi radiating chart and the notion of Bondi energy to rigorously study this question in maximal generality.

In 1957, at a conference at Chapel Hill, North Carolina, appealing to various mathematical tools developed by John Lighton Synge, A. Z. Petrov and André Lichnerowicz, Pirani explained more clearly than had previously been possible the central role played by the Riemann tensor and in particular the tidal tensor in general relativity.[4] He gave the first correct description of the relative (tidal) acceleration of initially mutually static test particles which encounter a sinusoidal gravitational plane wave.

21.2.3   Feynman's argument

Later in the Chapel Hill conference, Richard Feynman — who had insisted on registering under a pseudonym to express his disdain for the contemporary state of gravitational physics — used Pirani's description to point out that a passing gravitational wave should in principle cause a bead on a stick (oriented transversely to the direction of propagation of the wave) to slide back and forth, thus heating the bead and the stick by friction. This heating, said Feynman, showed that the wave did indeed impart energy to the bead and stick system, so it must indeed transport energy, contrary to the view expressed in 1955 by Rosen.

In two 1957 papers, Bondi and (separately) Joseph Weber and John Archibald Wheeler used this bead argument to present detailed refutations of Rosen's argument.[5][6]

21.2.4   Rosen's final views

Nathan Rosen continued to argue as late as the 1970s, on the basis of a supposed paradox involving the radiation reaction, that gravitational radiation is not in fact predicted by general relativity. His arguments were generally regarded as invalid, but in any case the sticky bead argument had by then long since convinced other physicists of the reality of the prediction of gravitational radiation.

21.3   See also

- Dashpot, of which the sticky-bead device is a variant.

- Monochromatic electromagnetic plane wave and monochromatic gravitational plane wave, for a modern account of two exact solutions which should clarify the point which confused Einstein and Rosen in 1936.

- pp-wave spacetime, for the Brinkmann gravitational wave solutions.

- Gravitational plane wave, for the Baldwin-Jeffery gravitational plane wave solutions.

- Brinkmann coordinates and Rosen coordinates for the two coordinate charts.

- Beck vacuums, for the Beck or Einstein-Rosen family of vacuum solutions.

- Cooperstock's Energy Localization Hypothesis, for a conflicting hypothesis (potentially, therefore, a test).

21.4   Notes

[1] Preskill, John and Kip S. Thorne. Foreword to *Feynman Lectures On Gravitation*. Feynman et al. (Westview Press; 1st ed. (June 20, 2002) p. xxv-xxvi.Link PDF

[2] DeWitt, Cecile M. (1957). Conference on the Role of Gravitation in Physics at the University of North Carolina, Chapel Hill, March 1957; WADC Technical Report 57-216 (Wright Air Development Center, Air Research and Development Command, United States Air Force, Wright Patterson Air Force Base, Ohio).

[3] Preskill, John and Kip S. Thorne. Forward to *Feynman Lectures On Gravitation*. Feynman et al. (Westview Press; 1st ed. (June 20, 2002) p. xxv-xxvi.

[4] Pirani, Felix A. E. (1957). "Invariant formulation of gravitational radiation theory". *Phys. Rev.* **105** (3): 1089–1099. Bibcode:1957PhRv..105.1089P. doi:10.1103/PhysRev.105.1089.

[5]Bondi, Hermann (1957). "Plane gravitational waves in general relativity".*Nature***179**(4569): 1072–1073.Bibcode:1957Natur. doi:10.1038/1791072a0.

[6] Weber, Joseph; and Wheeler, John Archibald (1957). "Reality of the cylindrical gravitational waves of Einstein and Rosen". *Rev. Mod. Phys.* **29** (3): 509–515. Bibcode:1957RvMP...29..509W. doi:10.1103/RevModPhys.29.509.

21.5 References

• Kennefick, Daniel (2005). "Einstein versus the Physical Review".*Physics Today***48**(9): 43.Bibcode:2005PhT....58 doi:10.1063/1.2117822. See also the on-line version

• Kennefick, Daniel, Controversies in the History of the Radiation Reaction problem in General Relativity

• Rosen, Nathan (1937). "Plane polarized waves in the general theory of relativity". *Phys. Z. Sowjetunion* **12**: 366–372.

• Einstein, Albert; and Rosen, Nathan (1937). "On gravitational waves".*J. Franklin Inst.***223**: 43–54.Bibcode:1937 doi:10.1016/S0016-0032(37)90583-0.

• Baldwin, O. R.; and Jeffery, G. B. (1926). "The relativity theory of plane waves". *Proc. Roy. Soc. Lond. A* **111** (757): 95–104. Bibcode:1926RSPSA.111...95B. doi:10.1098/rspa.1926.0051.

• Beck, Guido (1925). "Zur Theorie binärer Gravitationsfelder".*Z. Für Physik***33**: 713–738.Bibcode:1925ZPhy... doi:10.1007/BF01328358.

• H. W. Brinkmann (1925). "Einstein spaces which are mapped conformally on each other". *Math. Ann.* **18**: 119. doi:10.1007/BF01208647.

• Eddington, Arthur Stanley (1922). "The propagation of gravitational waves". *Proc. Roy. Soc. Lond. A* **102** (716): 268–282. Bibcode:1922RSPSA.102..268E. doi:10.1098/rspa.1922.0085.

• Einstein, Albert (1918). "Über Gravitationswellen". *Königlich Preussische Akademie der Wissenschaften Berlin Sitzungberichte*: 154–167.

21.6 Text and image sources, contributors, and licenses

21.6.1 Text

Catslash, Pcp071098, Bubba hotep, First Harmonic, Allstarecho, LorenzoB, Kornfan71, Davidjcmorris, Keith D, R'n'B, Rrostrom, Yonidebot, Tgebbie, Jpwest, Migran, Александр Сигачёв, Austin512, Novis-M, Tarotcards, Rominandreu, Wesino, DorganBot, Epistemenical, Sheliak, VolkovBot, Svmich, Sporti, Craigheinke, TXiKiBoT, MusicScience, Anonymous Dissident, Michael H 34, Broadbot, SwordSmurf, James McBride, Kbrose, Biscuittin, SieBot, Hertz1888, Csmart287, Wing gundam, Zbvhs, Csblack, Mimihitam, Jdaloner, RMB1987, Duae Quartunciae, Anchor Link Bot, Wyattmj, Martarius, GarbagEcol, ClueBot, The Thing That Should Not Be, Niceguyedc, Agge1000, ChandlerMapBot, I am a violinist, Excirial, Homonihilis, Nymf, Alexbot, Jefflayman, SolomonFreer, PixelBot, Bob108, Telekenesis, Tnxman307, Mastertek, Natty sci~enwiki, BOTarate, Panos84, Aitias, Nakomaru, Jonverve, DumZiBoT, BarretB, XLinkBot, DCCougar, BodhisattvaBot, Gwark, Dthomsen8, ErgoSum88, Ich42, Addbot, Dryphi, DOI bot, Ronhjones, Chotabe, Ka Faraq Gatri, Proxima Centauri, Ehrenkater, Astronorte, Lightbot, Zorrobot, Ben Ben, Luckas-bot, Yobot, Ptbotgourou, Legobot II, Aldebaran66, Amble, Wireader, Azcolvin429, AnomieBOT, Stuffed cat, Captain Quirk, Hunnjazal, Citation bot, Xqbot, Plastadity, Seb.mag, Nnivi, Cydelin, Srich32977, Lithopsian, J04n, GrouchoBot, EqualMusic, Frankie0607, Kyng, Amaury, Mnmngb, Bigger digger, Fotaun, CES1596, GliderMaven, Nagualdesign, FrescoBot, LucienBOT, Paine Ellsworth, Binrdow, Citation bot 2, HamburgerRadio, Citation bot 1, HRoestBot, MoonGirl78, Jonesey95, Tom.Reding, Lithium cyanide, Pmokeefe, RedBot, IVAN3MAN, RockSolidCosmo, TobeBot, Trappist the monk, Comet Tuttle, Michael9422, LI995, Earthandmoon, Tbhotch, Marie Poise, Wikiborg4711, Siranmichel, DexDor, Cwsavage78, Mathewsyriac, EmausBot, WikitanvirBot, Immunize, Quantanew, GoingBatty, Snorgway, Italia2006, Grondilu, ZéroBot, Medeis, Quondum, AManWithNoPlan, Miguelzuma, Iiar, Pumpkinking0192, Tbgriswold, Hang Li Po, ChiZeroOne, DASHBotAV, ClueBot NG, Ulflund, Factorial8, Helpful Pixie Bot, Bibcode Bot, BG19bot, Omegafold, AvocatoBot, Socal212, Ninney, Altaïr, Natalia.missana, Sparkie82, Fivemusketeers, U-95, ChrisGualtieri, JYBot, Dexbot, Neicdk, Manjolis, LightandDark2000, Antunesi, Reatlas, Rfassbind, User74~enwiki, Qmgsobserver, Praemonitus, Zlelik2000, OxygenBlueDansk, AbiLtoCen, Johndric Valdez, Exoplanetaryscience, Jlarks73, Monkbot, Filedelinkerbot, Falcon9v1.1, Unatnas1986, Trackteur, Werzaz, Anthul, SkyFlubbler, Samoniel1, Tullyojr, SwagYolo420ilovethis, Tetra quark, Anand2202, GeneralizationsAreBad, Freakcrane870, Feelthhis, Outedexits and Anonymous: 292

- **Graviton** *Source:* https://en.wikipedia.org/wiki/Graviton?oldid=677652863 *Contributors:* CYD, Bryan Derksen, Timo Honkasalo, XJaM, Fubar Obfusco, Maury Markowitz, Kaczor~enwiki, Jketola, TakuyaMurata, Eric119, Looxix~enwiki, Glenn, Cyan, Wooster, Charles Matthews, Timwi, Wik, BenRG, Donarreiskoffer, Scott McNay, Stephan Schulz, Arkuat, Chris Roy, Merovingian, Davidl9999, Giftlite, Xerxes314, Jason Quinn, Matt Crypto, CryptoDerk, RetiredUser2, Icairns, Zfr, Lumidek, Ukexpat, Urvabara, Discospinster, Pjacobi, Vapour, Brian0918, El C, Joanjoc~enwiki, Dalf, Army1987, Mpvdm, La goutte de pluie, Physicistjedi, Daniel Arteaga~enwiki, Zenosparadox, Dethtron5000, Keenan Pepper, Viridian, SidP, Falcorian, Skeejay, Simetrical, Dr Archeville, Mpatel, Kyleca, Tmassey, Christopher Thomas, Tevatron~enwiki, Kbdank71, Nightscream, Koavf, Mike Peel, Ems57fcva, FlaBot, RexNL, Chobot, DVdm, Roboto de Ajvol, Spacepotato, Anonymous editor, SnoopY~enwiki, Salsb, Bachrach44, Hyperbrand, NickBush24, Pnrj, RL0919, EEMIV, IslandGyrl, Bota47, C h fleming, Petri Krohn, Mario23, Alias Flood, Tim314, Teply, GrinBot~enwiki, SmackBot, Amcbride, Melchoir, Eskimbot, Gilliam, Skizzik, Timneu22, Complexica, Villarinho, Colonies Chris, Vladislav, Chlewbot, Xyzzyplugh, Jmnbatista, Fuhghettaboutit, Sadi Carnot, Yevgeny Kats, TenPoundHammer, Lambiam, Zaphraud, JorisvS, Mr Stephen, Ramuman, Quasar Jarosz, Lottamiata, Firewall62, Kurtan~enwiki, CmdrObot, BeenAroundAWhile, WeggeBot, Shultz IV, UncleBubba, Michael C Price, Anthmoo, Thijs!bot, Epbr123, Headbomb, KevinS06, Opelio, Spartaz, JAnDbot, Xoneca, SHCarter, Pikazilla, Robin S, STBot, Kostisl, J.delanoy, Tarotcards, Coppertwig, Wesino, Sava ankit2006, Tygrrr, Idioma-bot, Sheliak, JoAnneThrax, TXiKiBoT, WilliamSommerwerck, Hqb, Anonymous Dissident, Antixt, SieBot, Flyer22 Reborn, Henry Delforn (old), ClueBot, Ergn, Darkicebot, DenverRedhead, Addbot, Eric Drexler, Uruk2008, DOI bot, BrianBop, PJonDevelopment, F Notebook, Legobot, Picturesofnothing, Dov Henis, Alfredschrader, Eric-Wester, AnomieBOT, VanishedUser sdu9aya9fasdsopa, Jim1138, Materialscientist, Citation bot, Tomflaherty, ProtectionTaggingBot, Waleswatcher, FrescoBot, Juto20, LucienBOT, Paine Ellsworth, I dream of horses, Tom.Reding, RedBot, Omar.tigereyes, IVAN3MAN, Ashish.kotwal, Michael9422, D0wnfalle, EmausBot, Octaazacubane, 8digits, Slightsmile, K6ka, Thecheesykid, User10 5, Rcsprinter123, Orbjeeples, Puffin, Herk1955, ClueBot NG, Raidr, Masssly, Helpful Pixie Bot, Bibcode Bot, BG19bot, Shapoopy178, ServiceAT, PhnomPencil, Trevayne08, Brainssturm, Tjamcclain2, ChrisGualtieri, Ariscod, TheUyulala, LightandDark2000, Jessybun, Makecatbot, Kryomaxim, JRYon, Andyhowlett, Mark viking, Yorsh07, CensoredScribe, WPratiwi, Monkbot, Bryan Paul Senior, Dr.Begich, Nompynuthead, Jacobflarsen and Anonymous: 196

- **Gravitoelectromagnetism** *Source:* https://en.wikipedia.org/wiki/Gravitoelectromagnetism?oldid=677033222 *Contributors:* Derek Ross, Bryan Derksen, The Anome, Glenn, Reddi, Altenmann, Wolfkeeper, Mooquackwooftweetmeow, Beland, Rich Farmbrough, Pjacobi, ArnoldReinhold, Paul August, Bender235, Bookofjude, Rbj, I9Q79oL78KiL0QTFHgyc, Slicky, Pearle, Keenan Pepper, RJFJR, Ahazred8, GregorB, Qwertyus, Rjwilmsi, John Baez, Fresheneesz, BradBeattie, Jaraalbe, YurikBot, Hillman, Prime Entelechy, Spike Wilbury, SamuelRiv, 2over0, Josh3580, Petri Krohn, Geoffrey.landis, Nixer, SmackBot, Tom Lougheed, InverseHypercube, Vald, Mbset, Betacommand, Mirokado, Silly rabbit, Vladislav, John, Pthag, JorisvS, Cadaeib, Tawkerbot2, JRSpriggs, Chetvorno, CmdrObot, Vyznev Xnebara, Nicolas wolfwood, Cydebot, DumbBOT, Headbomb, WVhybrid, I do not exist, D.H, Blarrrgy, Peter Harriman, Arch dude, Pervect, Pixel ;-), R'n'B, Lantonov, Tcisco, Aervanath, Sheliak, Flyingidiot, Red Act, Antixt, Judgeking, Dmcq, Wing gundam, Dyeote, JerzyTarasiuk, Fedosin, The Thing That Should Not Be, VQuakr, Nike787, Djily, XLinkBot, PauloHelene, Addbot, Eric Drexler, DOI bot, Melab-1, Debresser, Favonian, Krano, Yobot, Kilom691, AnomieBOT, Citation bot, LilHelpa, Waleswatcher, Mnmngb, Carlog3, Originalwana, Citation bot 2, Citation bot 1, Tom.Reding, Quondum, Gyrogravitation, Maschen, ClueBot NG, Aacke, Ant.acke, Helpful Pixie Bot, Bibcode Bot, Manuelfeliz, Waitedavid137, Anto-nio 354, ChrisGualtieri, Dexbot, MiceEater, Elenceq and Anonymous: 96

- **Gravitational-wave observatory** *Source:* https://en.wikipedia.org/wiki/Gravitational-wave_observatory?oldid=687588193 *Contributors:* The Anome, William M. Connolley, Bearcat, Carnildo, Jason Quinn, RayTomes, Mpatel, Drbogdan, Rjwilmsi, Beanyk, SmackBot, Dicklyon, MystRivenExile, Thetrick, Myasuda, BobQQ, Markus Pössel, C0sbysweater, Jmcw37, Shiraun, Oshwah, Ng.j, McM.bot, MCTales, Martarius, DragonBot, Djr32, SchreiberBike, Johnuniq, TimothyRias, Addbot, Fieldday-sunday, SpBot, Zorrobot, Yobot, AnomieBOT, Citation bot, Xqbot, The Evil IP address, Originalwana, Citation bot 1, HRoestBot, Jonesey95, Tom.Reding, Henrysting, Rouvilleforever, Octaazacubane, Hhhippo, Quondum, Lsfinn, ChuispastonBot, RockMagnetist, Mikhail Ryazanov, ClueBot NG, GSZaum, Bibcode Bot, Negativecharge, Prasadkedar1997, Malyszkz, BattyBot, Dexbot, Pdecalculus, Monkbot, Popppppppy, Beckmanrj.18, Feelthhis and Anonymous: 19

- **LIGO** *Source:* https://en.wikipedia.org/wiki/LIGO?oldid=684088606 *Contributors:* AxelBoldt, Matthew Woodcraft, Bryan Derksen, Stevertigo, Skysmith, Whkoh, Wikiborg, Northgrove, RedWolf, Vuara, Lethe, Herbee, Wwoods, Curps, Peter Ellis, Mmm~enwiki, LucasVB, Lumidek, Urhixidur, TobinFricke, Diagonalfish, Discospinster, Brianhe, Wk muriithi, David Schaich, PhilipNeustrom, Bender235, Alereon, A2Kafir, Keenan Pepper, Apoc2400, Gene Nygaard, Alai, Linas, Mpatel, Dhs, TotoBaggins, Marudubshinki, Tevatron~enwiki, Rnt20, Rjwilmsi,

TheRingess, Mike Peel, Brighterorange, The wub, Burris, Williamborg, FlaBot, GangofOne, Amaurea, Wavelength, Wormholio, Hillman, JarrahTree, RussBot, Eleassar, Długosz, Gront, Philbull, Rayc, Zunaid, Arthur Rubin, Johnpseudo, Caco de vidro, John Broughton, SmackBot, Scorpiona, Herbm, Hmains, Skizzik, JohnWayne, Kearby, DavidBailey, Big Smooth, MOBle, Achoo5000, Mssgill, Cydebot, BobQQ, Shitok, Quibik, Sobreira, Cool Blue, Dawnseeker2000, Escarbot, JustinGarofoli, Magioladitis, RogierBrussee, Hroðulf, Bongwarrior, Swpb, Mamojama, Nvf, Pushnell, BatteryIncluded, Enquire, DerHexer, Stuver, MartinBot, Adavidb, Plasticup, Wesino, Potatoswatter, TXiKiBoT, Falcon8765, Klappspatier, Csblack, Vbond, Kumioko (renamed), Anchor Link Bot, Ascidian, ClueBot, Trojancowboy, Marjaliisa, Agge1000, Cavaglia~enwiki, Ddgutierrez, Jovianeye, BOinenglish, Kbdankbot, Addbot, Willking1979, Cuaxdon, Ufgatorman, Lightbot, Marhorr, Legobot, Luckas-bot, Yobot, AnomieBOT, Galoubet, Xqbot, Miguel in Portugal, GrouchoBot, Omnipaedista, Kismalac, OgreBot, Diego diaz espinoza, Foobarnix, Ale And Quail, Wikiborg4711, EmausBot, John of Reading, ZéroBot, Zach444, Jodoka, Hardywu, Verbithium, ClueBot NG, Morlockdigger, Helpful Pixie Bot, Bibcode Bot, BG19bot, Shawn Worthington Laser Plasma, Samwalton9, ChrisGualtieri, Joeinwiki, Monkbot, Donosauro, Dhr pas ca, Dick shitter, AtomicFountain and Anonymous: 122

- **Virgo interferometer** *Source:* https://en.wikipedia.org/wiki/Virgo_interferometer?oldid=645014278 *Contributors:* Hooperbloob, Keenan Pepper, Eubot, RussBot, Chesnok, Attilios, SmackBot, Melchoir, RevenDS, Silly rabbit, Wwagner, ShelfSkewed, BobQQ, Mglovesfun, Thijs!bot, Thecabinet, Magioladitis, The Anomebot2, Filippomdp, FKmailliW, Klappspatier, Japs it, Alexbot, Coinmanj, WikHead, Addbot, Bounty braveheart, DirlBot, MondalorBot, Alienautic, Cnwilliams, Extra999, Wonderwulf, Monkbot and Anonymous: 10

- **Evolved Laser Interferometer Space Antenna** *Source:* https://en.wikipedia.org/wiki/Evolved_Laser_Interferometer_Space_Antenna?oldid= 684037739 *Contributors:* SimonP, Heron, Docu, Dysprosia, Cos111, Sanders muc, Seth Ilys, Lupin, Curps, Maver1ck, Peter Ellis, Lumidek, Oliver Jennrich, Rich Farmbrough, Slipstream, Mr. Billion, BrokenSegue, Shenme, Jag123, Hooperbloob, Quaoar, Jumbuck, Grutness, Alansohn, Keenan Pepper, Snowolf, Falcorian, Mpatel, Tabletop, Rnt20, Kieran A. Carroll, Nanite, ConradKilroy, Tawker, Mike Peel, Vegaswikian, FlaBot, Eubot, Gurch, Chobot, Wjfox2005, Neum, Beanyk, 2over0, SmackBot, Cattus, MalafayaBot, Silly rabbit, Migatu~enwiki, Meithan, Olaf Davis, IntrigueBlue, Cydebot, BobQQ, Quibik, Martin Hogbin, Elpatg, John254, Dawnseeker2000, Jason-hamdaoui, .anacondabot, Magioladitis, BatteryIncluded, LorenzoB, Eschnett, Wikianon, R'n'B, Hans Dunkelberg, Jotempe, Ohms law, Steel1943, RJASE1, Idioma-bot, Funandtrvl, Anonymous Dissident, Retiono Virginian, Ng.j, Jackfork, Mazarin07, Louepower, Klappspatier, AlleborgoBot, Pditmar~enwiki, Quest for Truth, Astrobit, Tlamatini, MBK004, ClueBot, Trojancowboy, U5K0, Auntof6, Jo Lorib, Djr32, CohesionBot, HumphreyW, TimothyRias, Addbot, 37ophiuchi, 84user, Yobot, AnomieBOT, Shalley303, JackieBot, Csigabi, Xqbot, Erud, Gap9551, Lupettus, LucienBOT, D'ohBot, DrilBot, Tom.Reding, Gryllida, Olga Vovk, Earthandmoon, RjwilmsiBot, Akesich, EmausBot, Pippo skaio, CrimsonBot, Yiosie2356, SkywalkerPL, Favata, ChiZeroOne, Scienceface, El Roih, Theaitetos, Theopolisme, Bibcode Bot, BG19bot, Xlicolts613, BattyBot, ChrisGualtieri, JRiegerMM, Monkbot, Anuvarshanw, HelicalLightbulb, Kashish Arora and Anonymous: 80

- **Linearized gravity** *Source:* https://en.wikipedia.org/wiki/Linearized_gravity?oldid=666746701 *Contributors:* Phys, Lumidek, Pearle, Pol098, Mpatel, Eyu100, Archelon, Geoffrey.landis, KasugaHuang, Sardanaphalus, SmackBot, Commander Keane bot, Radagast83, JRSpriggs, Spiffyzha, Nick Number, Liquid-aim-bot, Alphachimpbot, Spartaz, Sikory, R'n'B, Sheliak, Red Act, Barkeep, 1ForTheMoney, Addbot, Citation bot, Carlog3, Ribashka, Laurifer, Quondum, Vanished user lt94ma34le12, Frinthruit and Anonymous: 7

- **Quadrupole formula** *Source:* https://en.wikipedia.org/wiki/Quadrupole_formula?oldid=649019936 *Contributors:* TimothyRias and Tom.

- **SQUID** *Source:* https://en.wikipedia.org/wiki/SQUID?oldid=688341035 *Contributors:* TwoOneTwo, Rmhermen, William Avery, Tzartzam, Twilsonb, RTC, Collabi, Ahoerstemeier, Julesd, Glenn, 4lex, Maximus Rex, Finlay McWalter, Robbot, Rorro, Carnildo, Wizzy, DavidCary, Niteowlneils, Frencheigh, McE~enwiki, Finn-Zoltan, Gzuckier, Symmetry, Jkl, Rich Farmbrough, El C, Walkiped, Slicky, Mr2001, Ynhockey, Bucephalus, Gene Nygaard, Falcorian, Linas, Polyparadigm, Robert K S, GregorB, Dbutler1986, Eteq, Saperaud~enwiki, Rjwilmsi, Eubot, Chobot, Krishnavedala, YurikBot, David Woodward, Gaius Cornelius, DavidConrad, Tjarrett, Tony1, Zwobot, Scottfisher, Cspalletta, Tyrhinis, SmackBot, Xlez8057, Chaojoker, Chris the speller, Thumperward, Colonies Chris, Tsca.bot, Stevenmitchell, Lostart, Rolinator, Onionmon, DabMachine, Marcusl, Brendanlevy, Tawkerbot2, E goldobin, Zureks, N2e, Sahrin, Supremeknowledge, Dr.enh, Thijs!bot, Arcresu, Tirkfl, Oosh, 1-54-24, Luna Santin, JAnDbot, Skomorokh, Twarge, Ariel., Vzapf, Nono64, Flyguy0507, Rod57, Saryakhran, ArqMage, Holme053, LokiClock, TXiKiBoT, Mercurywoodrose, CoJaBo, Erikev, AstroNerd2000, 4RM0~enwiki, Rmn1791, Barriga.a.s, Chemawb, Trojancowboy, ChandlerMapBot, ParisianBlade, Dthomsen8, Addbot, Mr0t1633, RPHv, JGKlein, Lightbot, Luckas-bot, Yobot, Aldebaran66, AnomieBOT, Piano non troppo, Crystal whacker, Materialscientist, Citation bot, FrescoBot, Citation bot 1, Sheetals magnetics, I dream of horses, Talkirz, Blackbaud, SingingZombie, Bobby122, RjwilmsiBot, Hhhippo, ClueBot NG, David C Bailey, Xavier Thibault, Mailme.kpriyadharsan, Helpful Pixie Bot, Bibcode Bot, BG19bot, Adwaele, Dexbot, Angufan, Jharrism2, Fedora99, DAsia, HRDVinc and Anonymous: 102

- **MiniGrail** *Source:* https://en.wikipedia.org/wiki/MiniGrail?oldid=660944978 *Contributors:* ErikvDijk, D6, RJHall, Woohookitty, Bgwhite, SmackBot, Bluebot, A5b, Headbomb, Dr. Submillimeter, Leyo, Ng.j, SieBot, Addbot, Yobot, Citation bot, Citation bot 1, Tom.Reding, Timetraveler3.14, Helpful Pixie Bot, Bibcode Bot, Jordatech, Monkbot and Anonymous: 8

- **GEO600** *Source:* https://en.wikipedia.org/wiki/GEO600?oldid=660982388 *Contributors:* Derek Ross, Alfio, Zoicon5, Nurg, HorsePunchKid, Tdent, Sysy, Bender235, Viriditas, Keenan Pepper, Rjwilmsi, FlaBot, Hillman, Pawyilee, SmackBot, Rokfaith, Spireguy, James Esterline, Bigmantonyd, Vgy7ujm, DavidBailey, Gurm, Pavithran, Vyznev Xnebara, Pph~enwiki, Michael C Price, Martin Hogbin, The Anomebot2, Klappspatier, The Thing That Should Not Be, Alexbot, Alexey Muranov, Addbot, Tide rolls, Zorrobot, Luckas-bot, Yobot, Xqbot, Jkbw, Miguel in Portugal, Yeroc99, RibotBOT, SassoBot, Citation bot 1, Tom.Reding, Ale And Quail, Stephenfairhurst, Ripchip Bot, Akesich, Zueignung, Bibcode Bot, Hooh1, BattyBot, Monkbot and Anonymous: 24

- **TAMA 300** *Source:* https://en.wikipedia.org/wiki/TAMA_300?oldid=628897900 *Contributors:* Fukumoto, SmackBot, The Anomebot2, Ng.j, Klappspatier, Eingangskontrolle, Addbot, LucienBOT, D'ohBot, El Roih and Anonymous: 3

- **KAGRA** *Source:* https://en.wikipedia.org/wiki/KAGRA?oldid=660965487 *Contributors:* Sappe, Bender235, Fivemack, SmackBot, MarshBot, Yellowdesk, Sheliak, TXiKiBoT, Ng.j, Lightbot, Luckas-bot, Yobot, Citation bot 1, Tom.Reding, Earthandmoon, Zach444, El Roih, KLBot2, Bibcode Bot, Mpitkin, Monkbot and Anonymous: 5

- **Deci-hertz Interferometer Gravitational wave Observatory** *Source:* https://en.wikipedia.org/wiki/Deci-hertz_Interferometer_Gravitational_ wave_Observatory?oldid=686834660 *Contributors:* Bearcat, Wjfox2005, Cydebot, BobQQ, C0sbysweater, Fotaun, Tom.Reding, El Roih, Mr Sheep Measham, Bibcode Bot, Ninney, Hmainsbot1 and Anonymous: 1

- **Pp-wave spacetime** *Source:* https://en.wikipedia.org/wiki/Pp-wave_spacetime?oldid=666287121 *Contributors:* Michael Hardy, Topbanana, Giftlite, Rich Farmbrough, Pt, Worldtraveller, BRW, -Ril-, Mpatel, Rjwilmsi, Ligulem, Srleffler, Wavelength, Hillman, Ilmari Karonen, That Guy, From That Show!, Sardanaphalus, SmackBot, Argyll Lassie, Bluebot, Colonies Chris, Chalybs, Ligulembot, CmdrObot, R'n'B, Sheliak, Neparis, Addbot, Uruk2008, DOI bot, Citation bot, Carlog3, Molitorppd22, Citation bot 1, Mjmarkowitz, Math-grav, DavidWTian, Bibcode Bot, SJ Defender, Frinthruit and Anonymous: 7

- **Spin-flip** *Source:* https://en.wikipedia.org/wiki/Spin-flip?oldid=622990079 *Contributors:* Edward, Rjwilmsi, SmackBot, The Thing That Should Not Be, Michael.r.sabino, Addbot, Sriharsha.karnati, Yobot, X-shaped, Pickhorn, Alsimone, WillisTheRoundhead, J04n, Rbm astro, MagnusRobotFighter, Raul5001, BenzolBot, Zurich Astro and Anonymous: 14

- **Sticky bead argument***Source:*https://en.wikipedia.org/wiki/Sticky_bead_argument?oldid=684423028*Contributors:*Bryan Derksen,Miguel Edward, Kku, Phys, Mervyn, Y(J)S, Laurascudder, Arthena, Starwed, Rangek, GangofOne, Hillman, Bruguiea, Długosz, SmackBot, Mgreenbe, Jcbarr, Hve, Ligulembot, Tesseran, Writtenonsand, CmdrObot, Gregbard, Dougher, Lantonov, TeamZissou, Likebox, Hamiltondaniel, Perry-Tachett, DOI bot, TheAMmollusc, NOrbeck, Citation bot 1, Earthandmoon, Dewritech, Cogiati, Brandmeister, ClueBot NG, Bibcode Bot, Goedelite, BattyBot, FiredanceThroughTheNight, Beachdadair and Anonymous: 13

21.6.2 Images

- **File:Ambox_important.svg** *Source:* https://upload.wikimedia.org/wikipedia/commons/b/b4/Ambox_important.svg *License:* Public domain *Contributors:* Own work, based off of Image:Ambox scales.svg *Original artist:* Dsmurat (talk · contribs)

- **File:Ambox_wikify.svg** *Source:* https://upload.wikimedia.org/wikipedia/commons/e/e1/Ambox_wikify.svg *License:* Public domain *Contributors:* Own work *Original artist:* penubag

- **File:Atomic_Interferometry.ogv** *Source:* https://upload.wikimedia.org/wikipedia/commons/5/57/Atomic_Interferometry.ogv *License:* Public domain *Contributors:* Goddard Multimedia *Original artist:* NASA/Goddard Space Flight Center

- **File:Cmbr.svg** *Source:* https://upload.wikimedia.org/wikipedia/commons/c/cd/Cmbr.svg *License:* Public domain *Contributors:* Own work *Original artist:* Quantum Doughnut

- **File:Commons-logo.svg** *Source:* https://upload.wikimedia.org/wikipedia/en/4/4a/Commons-logo.svg *License:* ? *Contributors:* ? *Original artist:* ?

- **File:Crab_Nebula.jpg** *Source:* https://upload.wikimedia.org/wikipedia/commons/0/00/Crab_Nebula.jpg *License:* Public domain *Contributors:* HubbleSite: gallery, release. *Original artist:* NASA, ESA, J. Hester and A. Loll (Arizona State University)

- **File:DC_SQUID.svg** *Source:* https://upload.wikimedia.org/wikipedia/commons/4/47/DC_SQUID.svg *License:* CC BY-SA 3.0 *Contributors:* Own work *Original artist:* Miraceti

- **File:Edit-clear.svg** *Source:* https://upload.wikimedia.org/wikipedia/en/f/f2/Edit-clear.svg *License:* Public domain *Contributors:* The *Tango! Desktop Project.* *Original artist:*
  The people from the Tango! project. And according to the meta-data in the file, specifically: "Andreas Nilsson, and Jakub Steiner (although minimally)."

- **File:Energia_template.svg** *Source:* https://upload.wikimedia.org/wikipedia/commons/0/00/Energia_template.svg *License:* CC-BY-SA-3.0 *Contributors:* Own work *Original artist:* user:Urutseg

- **File:Folder_Hexagonal_Icon.svg** *Source:* https://upload.wikimedia.org/wikipedia/en/4/48/Folder_Hexagonal_Icon.svg *License:* Cc-by-sa-3.0 *Contributors:* ? *Original artist:* ?

- **File:Gnome-searchtool.svg** *Source:* https://upload.wikimedia.org/wikipedia/commons/1/1e/Gnome-searchtool.svg *License:* LGPL *Contributors:* http://ftp.gnome.org/pub/GNOME/sources/gnome-themes-extras/0.9/gnome-themes-extras-0.9.0.tar.gz *Original artist:* David Vignoni

- **File:Gravitational-wave_detector_sensitivities_and_astrophysical_gravitational-wave_sources.png** *Source:* https://upload.wikimedia.org/wikipedia/commons/a/af/Gravitational-wave_detector_sensitivities_and_astrophysical_gravitational-wave_sources.png *License:* CC BY-SA 1.0 *Contributors:* http://www.ast.cam.ac.uk/~{}rhc26/sources/ *Original artist:* Christopher Moore, Robert Cole and Christopher Berry

- **File:GravitationalWave_CrossPolarization.gif***Source:*https://upload.wikimedia.org/wikipedia/commons/b/b8/GravitationalWave_Cross gif *License:* Public domain *Contributors:* Transferred from en.wikipedia to Commons. Transfer was stated to be made by User:Tgr. *Original artist:* MOBle at English Wikipedia

- **File:GravitationalWave_PlusPolarization.gif***Source:*https://upload.wikimedia.org/wikipedia/commons/b/b8/GravitationalWave_Plus gif *License:* Public domain *Contributors:* Transferred from en.wikipedia to Commons. Transfer was stated to be made by User:Tgr. *Original artist:* MOBle at English Wikipedia

- **File:Gravitational_lens-full.jpg** *Source:* https://upload.wikimedia.org/wikipedia/commons/0/02/Gravitational_lens-full.jpg *License:* Public domain *Contributors:* ? *Original artist:* ?

- **File:Gravitomagnetic_field_due_to_angular_momentum.svg***Source:*https://upload.wikimedia.org/wikipedia/commons/0/0f/Gravitom field_due_to_angular_momentum.svg *License:* Public domain *Contributors:* Own work *Original artist:* Maschen

- **File:Gravity_Probe_B_Confirms_the_Existence_of_Gravitomagnetism.jpg** *Source:* https://upload.wikimedia.org/wikipedia/commons/9/9f/Gravity_Probe_B_Confirms_the_Existence_of_Gravitomagnetism.jpg *License:* Public domain *Contributors:* APOD *Original artist:* Gravity Probe B Team, Stanford, NASA

- **File:History_of_the_Universe.svg** *Source:* https://upload.wikimedia.org/wikipedia/commons/d/db/History_of_the_Universe.svg *License:* CC BY-SA 3.0 *Contributors:* Own work *Original artist:* Yinweichen

- **File:Horn_Antenna-in_Holmdel,_New_Jersey.jpeg***Source:*https://upload.wikimedia.org/wikipedia/commons/f/f7/Horn_Antenna-in_H 2C_New_Jersey.jpeg *License:* Public domain *Contributors:* Great Images in NASA Description *Original artist:* NASA

- **File:SQUID_IV.jpg** *Source:* https://upload.wikimedia.org/wikipedia/commons/7/71/SQUID_IV.jpg *License:* Public domain *Contributors:* Own work *Original artist:* Jan Olaf

- **File:SQUID_by_Zureks.jpg** *Source:* https://upload.wikimedia.org/wikipedia/commons/9/98/SQUID_by_Zureks.jpg *License:* CC BY-SA 3.0 *Contributors:* Own work *Original artist:* Zureks

- **File:Spacetime_curvature.png** *Source:* https://upload.wikimedia.org/wikipedia/commons/2/22/Spacetime_curvature.png *License:* CC-BY-SA-3.0 *Contributors:* ? *Original artist:* ?

- **File:SpinFlip.jpg** *Source:* https://upload.wikimedia.org/wikipedia/commons/5/51/SpinFlip.jpg *License:* Public domain *Contributors:* Transferred from en.wikipedia to Commons. *Original artist:* Pickhorn at English Wikipedia

- **File:Squid_prototype.jpg** *Source:* https://upload.wikimedia.org/wikipedia/commons/b/b9/Squid_prototype.jpg *License:* Public domain *Contributors:* from NASA, Stanford University. originally uploaded at en.wikip. original description page is/was here[1]. *Original artist:* original uploader en:User:Slicky

- **File:Squid_prototype2.jpg** *Source:* https://upload.wikimedia.org/wikipedia/commons/d/db/Squid_prototype2.jpg *License:* Public domain *Contributors:* from NASA, Stanford University. originally uploaded at en.wikip. original description page is/was here[1]. *Original artist:* original uploader en:User:Slicky

- **File:Stylised_Lithium_Atom.svg** *Source:* https://upload.wikimedia.org/wikipedia/commons/e/e1/Stylised_Lithium_Atom.svg *License:* CC-BY-SA-3.0 *Contributors:* based off of Image:Stylised Lithium Atom.png by Halfdan. *Original artist:* SVG by Indolences. Recoloring and ironing out some glitches done by Rainer Klute.

- **File:TAMA300-vacuum-pump-and-beam-duct-tunnel.jpg***Source:*https://upload.wikimedia.org/wikipedia/commons/8/88/TAMA300-jpg *License:* CC BY-SA 3.0 *Contributors:* Own work *Original artist:* Kestrel (<a href='//commons.wikimedia.org/wiki/User_talk:Kestrel' title='User talk:Kestrel'>talk</a>)

- **File:Text_document_with_red_question_mark.svg** *Source:* https://upload.wikimedia.org/wikipedia/commons/a/a4/Text_document_with_red_question_mark.svg *License:* Public domain *Contributors:* Created by bdesham with Inkscape; based upon Text-x-generic.svg from the Tango project. *Original artist:* Benjamin D. Esham (bdesham)

- **File:VIRGO_-_view_of_west_tube.jpg** *Source:* https://upload.wikimedia.org/wikipedia/commons/9/95/VIRGO_-_view_of_west_tube.jpg *License:* CC BY-SA 3.0 *Contributors:* Снимок автора *Original artist:* Сивцов Иван

- **File:Virgo_detector_sensitivity_curve.png***Source:*https://upload.wikimedia.org/wikipedia/commons/2/2c/Virgo_detector_sensitivity_curve .png*License:*CC BY-SA 3.0*Contributors:*http://www.ast.cam.ac.uk/~{}rhc26/sources/*Original artist:*Christopher Moore, Robert Cole and Christopher Berry

- **File:Wavy.gif** *Source:* https://upload.wikimedia.org/wikipedia/commons/b/b8/Wavy.gif *License:* Public domain *Contributors:* ? *Original artist:* ?

- **File:Wikinews-logo.svg** *Source:* https://upload.wikimedia.org/wikipedia/commons/2/24/Wikinews-logo.svg *License:* CC BY-SA 3.0 *Contributors:* This is a cropped version of Image:Wikinews-logo-en.png. *Original artist:* Vectorized by Simon 01:05, 2 August 2006 (UTC) Updated by Time3000 17 April 2007 to use official Wikinews colours and appear correctly on dark backgrounds. Originally uploaded by Simon.

- **File:Wikiquote-logo.svg** *Source:* https://upload.wikimedia.org/wikipedia/commons/f/fa/Wikiquote-logo.svg *License:* Public domain *Contributors:* ? *Original artist:* ?

- **File:Wikisource-logo.svg** *Source:* https://upload.wikimedia.org/wikipedia/commons/4/4c/Wikisource-logo.svg *License:* CC BY-SA 3.0 *Contributors:* Rei-artur *Original artist:* Nicholas Moreau

21.6.3 Content license

www.ingramcontent.com/pod-product-compliance
Lightning Source LLC
Chambersburg PA
CBHW081458170526
45166CB00008B/2469